Crossing the Cusp

Surviving the Edgar Cayce Pole Shift

The Bad News You Expect and
the Good News You Need

Crossing the Cusp

Surviving the Edgar Cayce Pole Shift

From the Publisher of The Kolbrin Bible

Your Own World Books
Nevada USA

CrossingTheCusp.com
YowBooks.com
Kolbrin.com

COPYRIGHT

Crossing the Cusp: Surviving the Edgar Cayce Pole Shift
Marshall Masters

No part of this book may be reproduced or transmitted in any form or by any means, graphic, electronic, or mechanical, including photocopying, recording, taping, or by any information storage retrieval system, without the written permission of the publisher.

Trade Paperback
First Edition — January 2011
©2011 Your Own World, Inc.
ISBN-13: 978-1-59772-180-6
ISBN-10: 1-59772-180-8
CrossingTheCusp.com

YOUR OWN WORLD BOOKS
an imprint of Your Own World, Inc.
Nevada USA
YowBooks.com
Kolbrin.com
SAN: 256-1646

Notices

Every effort has been made to make this book as complete and as accurate as possible and no warranty or fitness is implied. All of the information provided in this book is provided on an "as is" basis. The authors and the publisher shall not be liable or responsible to any person or entity with respect to any loss or damages arising from the information contained herein.

Trademarks

All terms mentioned in this book that are known to be trademarks or service marking have been capitalized. Your Own World, Inc. cannot attest to the accuracy of this information and the use of any term in this book should not be regarded as affecting the validity of any trademark or service mark.

Special Thanks To

- Edtior: Jacqueline Ramey
- Cover: Richard Shaw, www.pinlight.com
- Illustrations: Anthony Diecidue, www.artofant.com
- Avebury 2008 Aerial Photo: Eva-Marie Brekkestø

Dedication

Edgar Cayce (1877-1945)

Micah 6:8: *New International Version*
He has shown you, O mortal, what is good.
And what does the LORD require of you?
To act justly and to love mercy
and to walk humbly with your God.

Edgar showed us the way.

Table of Contents

Part 1 — The Bad News..........................1

1. The Big Picture..........................3
    We Are Born to Evolve..........................3
    Succeeding Where Past Generations Have Failed..........................4
    Global Catastrophe as the Engine of Human Evolution..........................6
    Catastrophism vs. Creationism and Darwinism..........................6
        Creationism..........................6
        Darwinism..........................7
        Catastrophism..........................7
        The Key Differences..........................8
    Darwinism as a Political Solution..........................8
    The Return of Catastrophism..........................9
    What This Means to You..........................10
    Surviving as an Independent Thinker..........................11

2. The Last Pole Shift..........................13
    The Deluge Story of Sisuda and Hanok..........................14
    What Triggered the Deluge?..........................20

3. The Trigger Event..........................21
    Crop Circles, Deep Time History, and Global Catastrophes..........................22
    Single Point of Truth (SPOT)..........................27
    The Struggle with Planted Disinformation..........................28
    Avebury Part 1 of 2..........................30
        July 15, 2008 – The First Part Appears..........................30
        An Asymmetrical Formation..........................31
        Winter Solstice Time Stamp..........................31
    Avebury Part 2 of 2..........................32
        July 22, 2008 – The Second Part Appears..........................33
        The Complete Formation..........................33
        Size Does Matter..........................34

| *The Minestrone*..35
| *Tramlines in the Formation*...36
| *The Avebury Binocular*...37
| The Ecliptic..38
| *Where?*...39
| Path of the Destroyer...40
| *A Candlelight Conundrum*..42
| *Pluto's Fate*..44
| *The Kozai Mechanism*...44
| *The Math Just Doesn't Add Up!*...45
| *The Path of Planet X*...46
| *The Whip*..47
| The Kill Zone...48
| *A Waning Crescent Moon*..49
| *Shelter by Day – Forage by Night*......................................49
| *Trigger Event*...51

4. The Dragon's Tail 53

What You Can't See Won't Hurt You..54
The Dust Tail..55
It Will Appear as a Smaller Sun...56
The Nature of the Beast...58
   The South Pole Telescope...58
   NibiruShock2012 and DNIr4808n Disclosure Videos............59
   What the Disclosure Videos Revealed....................................60
The Dragon's Tail..61
   Starting From Avebury..64
Earth's First Transit of the Dust Tail...65
   The Price of False Grandeur..66
The Last Flyby as Told by the Ancient Egyptians........................67
Meteorites, Algae Blooms, and Microcystin................................69

5. The Great Winnowing 73

Is a Pole Shift Actually Possible?...74
The Search for a Cusp Survival Paradigm....................................75
The Noah Paradigm...75
The Noah Dilemma...76
   A Righteous Man...77
   Choices and Consequences..77

- Defining Attributes of a Noah Paradigm...78
- Flyby and Pole Shift 2012 to 2014...79
- Phase One — 4th Qtr 2012 Through 2nd Qtr 2013...80
- Phase Two — 2nd Qtr 2013 Through 1st Qtr 2014...82
  - The Downside of Hopeful Assumptions...84
  - Up to the Bottom...84
  - Beware the False Grandeur...85
  - Phase Three — 2nd Qtr 2014 Through 4th Qtr 2014...85
- The Resonance Effect...87
  - Negative Feedback...87
  - Positive Feedback...87
  - Catastrophic Tipping Point...88
- Earth's Inevitable Tipping Point...88
- The Pole Shift...90
  - If the Moon Disappeared...91
  - A One-Two Punch...91
- The Greatest Single Cause of Death...92
  - Chaos Before the Tipping Point...92
  - Fear Is the Knife's Edge of Death...93
- The Great Winnowing Within The Cusp...93
  - A Cosmic Bad Trip...94
  - When the Great Winnowing Begins...94
  - Fate of the Weakest...95
  - The Forceful Fall Next...95
  - What the Forceful Fear...96
- Elites and Their Minions...96
- The Enlightenment...97

Part 2 — Crossing the Cusp...99

6. Welcome to Awareness...101
- A Future Time for You...101
- Look Within Yourself...102
- An Amazing Thing to Watch...102
- Why Weren't We Warned Earlier?...103
- The Green Revolution Sputters Out...104
- Starvation by Design...106

    The Third Man on the Match..106
    Truth or Self-Deluding Fantasy..107
    Position, Position, Position..108
    The Ghost of Y2K Past..110
    The Meek Shall Inherit the Earth..112
    Hope vs. The Elites..114

7. How the Meek Prevail..115

    Entropy...116
    Harmony...117
    Fear and Love..118
    Natural Flow of Consciousness..119
    Lifeworkers and The Great Winnowing..121
    The Accelerant Effect...123
    Survivor Triage..124
    No One Voice Is Omnipotent..124
    The Burnout Point...127
    What the Entity Showed Betty...127

8. The Enlightenment..133

    Why the Meek Have the Best Odds...134
    Déjà Vu..135
    Dreams...135
    What's to Gain...136
    What Can We Find?..137
    Intellectual vs. Experiencial Definitions...137
    My First Out-of-Body Experience...138
    Know Your Limits..139
    Out-of-Phase Experience...140
    Shared-Death Experiences..143
    The Survival Benefits of a Shared-Death Experience...................145
    Oh My Johnny..146
    Fear-based Expectations..147
    The Fear of Death...149
    Today Is a Good Day to Die..149
    The Ancient Right of Monarchs..150
    Taking the First Step...151
    Observation and Contemplation...152

    The Owls of Aesop's Fables..153
    Transformative Experiences...154

9. Transformation..157
    My First Transformative Experience....................................158
    Reading No. 1 — Your Destiny...161
    Reading No. 2 — You Are Not Alone....................................162
    Reading No. 3 — The Tree of Life..164
    Reading No. 4 — Fear and Ascension..................................166
    Reading No. 5 — Beware of Exploiters.................................168
    Reading No. 6 — Hold Your Resolve....................................169
    Reading No. 7 — Nemesis, Planet X and Our Sun...............171
    Reading No. 8 — Signs of the Pole Shift.............................172
    Reading No. 9 — Sharing with Others.................................174
    Reading No. 10 — Heat and Radiation................................175
    Reading No. 11 — Consciousness.......................................176
    Reading No. 12 — The Purpose of It All..............................179

Part 3 — The Good News..................181

10. We Can Do This..183
    The Bucket List..184
    Deep Blue...185
    Visualize the Goal..186
    Bunker Bunnies and Earth Pilgrims....................................187
    Nature Designed Us to Be Earth Pilgrims..........................189
    Buying a Hammer Visualization...191
    The Five Zones of Control...195
    Zone 1 — Personal..197
    Zone 2 — Encampment..202
    Zone 3 — Outer Perimeter..203
    Zone 4 — Observation and Foraging.................................206
    Zone 5 — Electronic and Supernormal...............................207

11. We Have Friends...211
    The Big Picture..213
    We the People...214

    A Pyrrhic Victory..215
    The Wisdom of Serapis Bey...216
    We're All Going Somewhere..217
    About and Why..217
    Great Danger...218
    In Chaos Is Opportunity..219
    Be in It for the Species..220
    The Special Children..221
    Where Do You Go From Here?..223

12. We Can Reprogram the Future..225
    What Is the Lesson Here?..226
    Night Flying..228
    The Project EarthSave48 Simulator...229
    The Trigger Event..231
    Intuitive to All Languages..231
    Galactic Intention..232
    Measurable Scientific Goals..233
    Leveraging Galactic Intention...233
    Intention Vortices...234
    A Game of Cosmic Eight-Ball...235
    EarthSave48 Global Event Zones..237
    The Optimal Participants..239
    Preparing the Families..240
    Project Volunteers and Contributors..242
    First Intention Vortex Event Simulation..244
    Tokyo: The Night of the Event...247
    Tokyo: The Event Begins..251

Alphabetical Index...261

Illustration Index

1: Cycle of Catastrophism..7
2: Hopi Prophesy Rock..11
3: The Kolbrin Bible...14
4: Edgar Cayce..14
5: Ark at Sea Level..15
6: Avebury - July 22, 2008..21
7: Cut to the Chase, Show #17...23
8: Cut to the Chase, Show #106...26
9: Cut to the Chase, Show #116...26
10: Planet X Forecast..29
11: First Layer - July 15, 2008..30
12: Types of Formations..31
13: Sun in Ophiuchus..32
14: Second Layer - July 22, 2008...33
15: First and Second Layers...34
16: Size of Formation..35
17: Harbinger Symbols..35
18: Tram Lines...36
19: Ecliptic Ring...37
20: Ecliptic Side View..38
21: Ecliptic View..39
22: Planet X Inbound..40
23: Planet X Inbound..41

24: Symbol for Planet X..42

25: Pluto Perturbed...44

26: 03-hale-bopp..45

27: Path of Planet X..47

28: December 2012...48

29: In the Kill Zone...49

30: The Worst Begins...50

31: Pole Shift Trigger..51

32: Tail and Coma...56

33: We See Two Suns..57

34: NibiruShock2012 - Jan/Feb 2008..59

35: DNIr4808n - Sep 2008..60

36: SPT Leaks Composite..61

37: Mother Shipton...62

38: Approaching the Tail..64

39: First Transit 2013...65

40: Planet X Reappears...66

41: Second Transit 2014..67

42: Planet X - Kolbrin Bible..68

43: Flyby and Pole Shift...80

44: First Phase 2012 - 2013...81

45: Second Phase 2013 - 2014..83

46: Third Phase 2014...86

47: Multiple Event Resonance...89

48: Pole Shift Event..90

49: 05-the-cusp..92

50: The Two Absolutes..118

51: Flow of Consciousness..120

52: The Great Winnowing..122

53: Marshall Masters (1976)...139

54: Marshall Masters (1975)...141

55: Zones of Control 196

56: Zone 1 — Personal...197

57: Yovy Suarez Jimenez...198

58: 1911 Colt .45 - Bob Suchke..200

59: Zone 2 — Encampment...202

60: Zone 3 — Outer Perimeter...203

61: Zone 4 — Observation...206

62: Zone 5 — Electronic+..208

63: Marshall Masters (2000)..226

64: Eye of the Vortex..249

65: Hawaii Land Venue...250

66: Intention Vortex..253

67: Form the Base..254

68: Jonn Lennon...257

Marshall's Motto

*Destiny finds those who listen,
and fate finds the rest.*

*So learn what you can learn,
do what you can do,
and never give up hope!*

Part 1
The Bad News

"Chance favors the
prepared mind."
—*Louis Pasteur*

1
The Big Picture

Awareness has brought you to this point and you understand, sense, or dream that life as we know it is about to end—that humanity is about to enter a terrible tribulation that will lead a small share of humanity to the doorsteps of a new future.

With a sense of impending loss we ask: Is this new future worth the suffering? To answer, this book gives you the bad news you expect and the good news you need.

We Are Born to Evolve

As we each come into this awareness, a sense of hopelessness inevitably tugs at our own ambitions of survival. Eventually we each ask ourselves: Do I really want to survive a time when the living envy the dead? This is a natural and proper response and one we must each come to peace with, in our own way and in our own time.

For the common man, there is nothing to justify such personal suffering. However, for humanity as a whole, consider the Dark Ages in Europe. They were called dark for good reasons, to include perpetual successions of feudal wars and plagues.

The worst was the Black Death of 1348 to 1350. Generally believed to be an outbreak of the bubonic plague, by some estimates it claimed 60 percent of Europe's population. It brought about an upheaval in which the religious, social, and economic institutions of the day controlled the masses.

Although it was a brutal turning point for humanity, the Black Death and subsequent outbreaks of the plague nonetheless shaped the beginnings of what our world has become today.

As a consequence of this tribulation, the seeds of the Renaissance (French for "rebirth") that spanned the 14th through 17th centuries in Europe were planted.

These periods of death and suffering were brutal for those forced to endure the agony. However, in chaos and suffering are the first glimmers of profound evolutionary changes, and there were many. Three that favored the common man are:

- **Independent thought.** Failing to explain or prevent the Black Death, the Church's iron-hand domination over free thought was undermined as it was helpless to explain, or prevent, the sickness. Consequently, people began looking elsewhere for answers and so began exploring the natural world, not with doctrine, but through their own powers of observation. Despite ruthless Church suppression, this trend endured with notable results.

- **Modern science and medicine.** This new and irrepressible curiosity about the workings of the natural world sowed the first seeds of modern science and technology as we have come know them. One particular benefit of this schism between Church doctrine and science was that it would later spark the beginnings of modern medicine.

- **Reduced exploitation.** Prior to the Black Death, elites benefitted from a vast pool of cheap and easily exploited labor. After the Black Death, elites found themselves competing with each other for significantly fewer workers. This eventually brought about new forms of automation and marginally better conditions for workers.

As brutal as these times were, during the Renaissance humanity arose like the magnificent Phoenix from the ashes of these tribulations.

As odd as it may sound, humanity is similar to tree species that cannot reproduce without the devastation of a forest fire. Therefore, whether we wish to participate in humanity's ongoing evolution or not, the universe is still insistent. We either evolve or step aside for species that will.

What recorded history shows us is that humanity has crossed this cusp of evolution before and has always managed to hold on. This time will be different, however, because the stakes are much higher than ever before.

Succeeding Where Past Generations Have Failed

The peoples of ancient times were swept away by what is approaching us yet again. Consequently, only the tiniest fragments of their history remain to explain their greatness as well as their failures.

It is why the ancients sent their knowledge forward in time to us, so we can succeed where they failed by crossing the cusp of the next coming global evolutionary event. This event will reorder life on this planet and shape human destiny from this point forward.

What is at stake is the very future of humanity itself. That we shall survive is a given. The question then becomes to what end. Do we evolve into a peaceful and enlightened stellar species, where our descendants populate the voids of space? Or, will we devolve?

If we abandon our humanity out of fear, we will drift back to a time such as the 5th century. This was before the advent of Greek philosophy, which has shaped our present view of the world.

Worse yet, we could bifurcate into two species, much like the *Eloi* and *Morlocks* depicted in H.G. Wells' classic, *The Time Machine*. All these things are possible unless we choose the probable thing and persist in making that thing flourish.

Unlike our ancestors, we can both endure and choose to cross the cusp of evolution to the beginning of a noble future for humanity. We can do this because we, of this time, possess the numbers, enlightenment, and science to shape a first-of-a-kind opportunity for humanity.

We can choose a new reason to endure and carry on, to hope, to be in it for the species. It shall be the deciding reason. After all other reasons for taking the next step have failed for those living in service to self, this reason will be in sight, yet beyond their grasp.

However, for those who live in service to others, this new reason will fuel their resolve to take the next step, and the one after that, and so on. In this regard, there are only two types of people: those who close their eyes and see a universe of possibilities and those who close their eyes and see only the back of their eyelids.

It is why those who live in service to self fixate on dark outcomes painted on the back of their eyelids and magnified by their own fears. Do not let their pronouncements undermine your resolve for they are but a vocal and disgruntled minority. No matter how dark the hour, when you close your eyes, you must see what they cannot. And what must you see?

You must see a universe of possibilities that draws you to the hope of a noble future—one in which our descendents lovingly honor our courage and our sacrifice. We can take those difficult but vital next steps forward even after all other reasons have failed us.

And so our descendents will celebrate our determination to persevere and they will number in the many billions, if not the hundreds of billions, and our seed will be scattered across the night sky in a magnificent web of life.

However, there are risks.

If we stumble as our ancestors did, we of this generation will condemn our species to the subjugation of slavery, slavery that will last for countless generations to come. Then, after our world has been so badly raped and brutalized that it is of no value to anyone, we will be given our freedom for whatever time remains.

Therefore, in 2012 and beyond, those who successfully cross the cusp shall not do so because of wealth, power, or whatever God they pray to. They will cross the cusp because they are the dauntless, the stouthearted, and the "meek" as the word was first defined by our ancestors.

It is for this very purpose that this book was written. In the process of surviving the terrible tribulations to come, those who endure and live beyond them will come to a profound understanding—one that forever changes them as well as the course of human history. Therefore, defining this process of change is the necessary first step in understanding it.

Global Catastrophe as the Engine of Human Evolution

Among Western thinkers, Creationism and Darwinism are the two dominant views of evolution on this planet. These views are at perpetual loggerheads, principally by design, and can never be reconciled.

However, there is a third view of evolution on this planet that combines what some would call the best concepts of both Creationism and Darwinism. The result is a more realistic view of how life evolves on our planet, and it is quickly gaining popularity.

It is called Catastrophism and it serves as the basis for this book. Therefore, a brief introduction is necessary.

Catastrophism vs. Creationism and Darwinism

One could fill a library with books about evolution and how we all came to be. For this reason, the purpose of this book is to illustrate certain differences among world views as they pertain to the coming global tribulation. First, let's define the terms.

Creationism

Based on the Genesis creation narrative in the Jewish *Torah*, Creationism dates the age of the universe to be less than 10,000 years old.

Divine intervention is the core of this theological view of evolution, where all things, including humans, are the creations of a supernatural being.

Remembered for its confrontation with Darwinism in the 1925 Scopes trial in the state of Tennessee (USA), Creationism may have won the legal battle, but it nonetheless lost the war.

Consequently, a recent variant called Intelligent Design represents an effort by Creationists to present a more mainstream, scientifically palatable, updated version of Creationism.

Darwinism

Now regarded as a loosely defined generic term, "Darwinism" was coined in 1860 by Thomas Henry Huxley in his review of Charles Darwin's book, *On the Origin of Species* (1859).

Generally speaking, Darwinism offers an explanation of evolution based on scientific naturalism.

Widely held to be an outshoot of Uniformitarianism, it maintains that all things continue on as they have since the beginning of the world.

It further shares the view of Gradualism that evolutionary changes occur in a process of gradual steps.

Catastrophism

Catastrophism as an explanation of evolution on this planet was first put forth by Baron Georges Léopold Chrétien Frédéric Dagobert Cuvier (1769-1832). In 1817, he published his groundbreaking work, *Le Règne Animal* (*The Animal Kingdom*).

It is important to note that until the advent of Darwin, Catastrophism was a popularly held view among scientists and for good reason.

Based on the scientifically accepted age of the universe, Catastrophism at its essence describes what is actually observable in the deep time-evidence history of the planet.

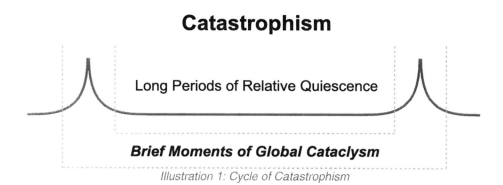
Illustration 1: Cycle of Catastrophism

Catastrophism tells us that evolution on this planet is primarily driven by long periods of relative quiescence punctuated by brief moments of global cataclysm.

Another version of Catastrophism, called Neocatastrophism, further maintains that this process is endemic to all Earth-like planets throughout the universe.

The Key Differences

For the purposes of this book, the three most essential comparisons among Creationism, Darwinism, and Catastrophism are:

- **Trigger events.** These are what actually cause the major evolutionary events that reorder life on the planet, not just for a few or even a single species but for the vast majority.

- **Divine intervention.** Is evolution driven solely by coincidental factors of natural processes? Or, is it in part or whole some form of intentional intervention by a deity, or even extraterrestrials for that matter?

- **Dating method.** Is the dating based on an experiential belief system or on independently verifiable observations? In other words, where Darwinism and Catastrophism measure the age of the universe in billions of years, Creationism measures it in thousands of years.

Catastrophism and Creationism agree in that they both maintain that brief moments of global cataclysm do occur (such as Noah's flood) and that they reorder life on the planet.

Herein was the unpalatable twist for 19th century scientists. While Catastrophism essentially rejects Creationism's divine intervention belief and its Biblical short-dating of the age of the universe, it does partly validate Creationism.

This is because Catastrophism's evolutionary cycle of long periods of relative quiescence punctuated by moments of global cataclysm is clearly in line with the Creationism theme.

Darwinism as a Political Solution

Because Catastrophism partially validates Creationism, 19th century scientists were delighted to find a way to firewall themselves altogether from the "unscientific" notions of religion.

Like an unhappy spouse searching for an expedient way to bolt from a miserable marriage, the rush to Darwinism had more to do with politics than science. This helps to explain why Darwin was able to circumvent the scientific peer-review *process*. The winds of political expediency were filling his sails.

Consequently, Darwinism gave science a way to completely divorce itself from any hint of Creationist theology and the Church as well.

As with all bad divorces, this final separation had its messy points as well.

Scientists had to ignore a crucial aspect of the Earth's deep time history so as to free themselves of the untenable perception that they had no choice but to validate a core concept of Creationism.

The result is that the awkward, but workable, truce between Catastrophism and Creationism was cast aside in favor of the deeply polarized debate that now reigns, and this bitter feud dominated the discussion until 1950.

Enter Immanuel Velikovsky.

The Return of Catastrophism

In 1950, a Russian-born independent scholar by the name of Immanuel Velikovsky (Иммануил Великовский) published his controversial book, *Worlds in Collision*.

Drawing on ancient mythology, folklore, and wisdom texts, including the *Holy Bible*, he dusted off the languishing concept of Catastrophism. He argued that the Earth had experienced catastrophic events due to close-contact encounters with Venus, Mars, and other planets.

A perfect fit for the ever-growing chasm between Creationism and Darwinism, Velikovsky's book caught on like wildfire.

Scientists went ballistic and in what is now regarded as science's shameful moment of medieval "McCarthyism," they largely succeeded in intimidating publishers, distributors, and book resellers so as to quash academic interest in Velikovsky's book, which remains popular to this day.

Even Carl Sagan, one of Velikovsky's most vocal critics, was dismayed by this abuse of free speech. In his "Cosmos" (1980) television series, he expressed his deep regret for the brutal manner in which Velikovsky had been treated by his peers.

Interestingly enough, though Sagan clearly disagreed with Velikovsky's conclusions on many levels, it could be argued that this clash intrigued Sagan with the concept of Catastrophism. In 1985, he co-authored *Comet* with Ann Druyan—a book that made the case for catastrophic cometary events as evolutionary triggers.

It is also important to note that in 1976 another independent researcher by the name of Zecharia Sitchin published his ever-popular work, *Twelfth Planet*. Much like Velikovsky, Sitchin drew upon his own translations and analysis of ancient Mesopotamian iconography and symbology.

According to his translations, a large body known to the ancients as Nibiru passes through the core of our solar system approximately every 3,600 years and with catastrophic results for the Earth.

While Velikovsky, Sitchin, Sagan, and others plowed the soil for a new look at Cuvier's original 1817 work on Catastrophism, it was Nobel Laureate Luis W. Alvarez and his son Walter who actually planted the seeds of Catastrophism's big comeback in 1980.

The Alvarez theory tells us that the age of the dinosaurs ended 65.95 million years ago after a huge asteroid struck an area called Chicxulub just off the coast of Mexico. Known as the K-T impact, this catastrophic event caused the extinction of the dinosaurs, which in turn made the age of man possible.

Now widely taught in schools, early adoption of the K-T impact was slow, and many in the scientific quarters still hotly challenge the theory as the sole source of the extinction.

Although the K-T impact may not have been the sole cause of the extinction of the dinosaurs, it certainly was a major trigger event, if not the principal trigger event.

What This Means to You

As the reader of this book, you are no doubt asking yourself: What does all this have to do with the subject of this book, and my own survival for that matter? Here's the answer.

This book is based on the concept of Catastrophism. It explains that while humanity is now the dominant species on this planet, our position is about to change.

Within a matter of years, we will bear witness to the end of what has been a long period of evolutionary quiescence—one that is, in fact, already sputtering out.

This in turn is setting us up for the key catastrophic trigger event that will set in motion a series of natural and manmade disasters that will reorder life on this planet once again. Granted, this is a terribly difficult idea to accept as our shortsighted, material societies simply leave us unprepared for it.

In fact, not one modern religious, political, or economic paradigm has ever survived a global cataclysm. Rather, what we really have is a replay of the Black Death described earlier.

Ergo, every *ology* and *ism* that presently defines our modern world is a long shot at best in terms of getting us through what awaits us around the bend. Furthermore, while ruling institutions profess noble intentions of harmony and coexistence, the simple fact is that greed and subjugation typically win the day.

Consequently, we must accept the fact that many of our present unsustainable paradigms will fail and with Titanic results. This of course begs the question: Does anyone have a proven track record? Yes!

The paradigms of indigenous peoples such as the Aborigines of Australia and the Hopi of the southwestern US have a solid track record of surviving global catastrophes.

Hopi Prophesy Rock Near Oraibi, Arizona USA

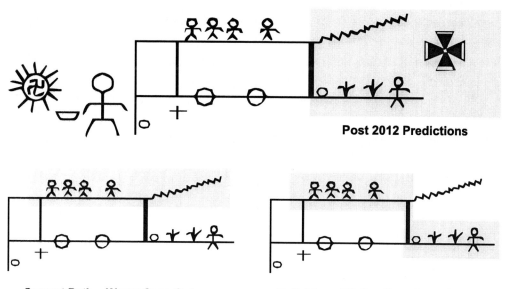

Illustration 2: Hopi Prophesy Rock

This is why their present-day descendants are preparing for the coming global catastrophe. They've taken their folklore and prophecies to heart as the harbingers foretold by their ancestors now come to pass.

Does this mean that all of us who walk the material path of a consumer society should cast aside our own beliefs and adopt those of the Aborigines, Hopi, and others?

While a few have, the answer is that this is not necessary for the vast majority of us. This, of course, assumes that we are willing to recognize the need to be independent thinkers—not the talk-the-talk kind, but the walk-the-walk kind.

Surviving as an Independent Thinker

The whole point of comparing Catastrophism with Darwinism and Creationism is that this insoluble debate is fueled by political and economic interests—not observable science.

- On one hand, Creationists completely dismiss scientific observations with theological blinders.
- On the other, Darwinists are so intent on divorcing themselves from theology that they've thrown a critical component of evolution under the bus.

Consequently, the only thing the two now have in common is their medieval treatment of anyone proposing an idea that is deemed to be heretical to their own canons.

So, here is the truth of it: The universe couldn't care less who wins, so why should we? The universe is about survival—that of ourselves, our loved ones, and our species.

It is why Darwinism and Creationism continue their duel, like a contentious parody of the current divide in American politics.

Even though Republicans and Democrats dominate the political dialogue, the pivotal segment of the electorate is independents and they're weary of the acrimony and false promises. Like Velikovsky and Sitchin, independent voters refuse to swear allegiance to one side or another. Rather, they follow their own interests independent of others.

Therefore, approaching the concept of Catastrophism, which is the basis of this book, requires an independent mind.

Listen to everybody and everything, but then decide in your own way and in your own time, as this is the vital first step in surviving the coming global cataclysm. This is because the only truth that matters is the truth that resonates within you. And that only happens when you and you alone, put it there.

All that being said, this book is simply the truth as it resonates with the author, and nothing more. Use it as you will, but if you really do want to survive the coming global tribulation along with your loved ones, you must being to find your own truth—and soon!

2

The Last Pole Shift

During his third and last voyage on the HMS Resolution, English explorer Captain James Cook made landfall in Hawaii in January 1778 at Waimea Harbor, Kauai, on the big island of Hawaii.

Today, tourists are often regaled with the story of how Cook was killed by Hawaiian villagers in 1779 while trying to launch his boats off the beach to escape a crisis he helped create. The one part of the story that always brings a sparkle to a tour guide's eyes is the story of Noah and the flood.

As was the custom of the day, Cook was introducing the Hawaiians to the stories of the *Holy Bible*. As the story of Noah's flood comes early in Genesis, it came up quickly.

To Cook's surprise, the Hawaiians told him they already knew the story and that the only difference between their version and his was in the spelling and pronunciation of Noah's name.

Cook spelled it "Noah" whereas the Hawaiians simply spelled it "Noa," without the "h." Other than that, the two accounts were very similar.

This bit of history highlights the fact that there could be as many as 500 different deluge accounts in the various ancient folklore and wisdom texts of cultures all across the globe.

One of the most scientifically prescient accounts is found in the *The Kolbrin Bible*. In it, we see an alternate account of Noah and the flood as told by the ancient Egyptians. Why look beyond the Genesis story for alternate accounts of what is widely known as the Great Deluge?

Today, we understand much more about our world and how it works. Ergo, many wonder if there was actually enough water on the surface of the Earth to flood all the land masses during the time of Noah.

A few may also wonder whether the Great Deluge was one component of a much greater event like a pole shift. If this is true, then it stands to reason it could happen again, and the possibility of another global catastrophe of Biblical proportions may lie in our future.

Illustration 3: The Kolbrin Bible

The premise of this book is that we shall experience such an event and that it will be an evolutionary event for humanity as we cross the cusp during the next pole shift event. This is the very event predicated by Edgar Cayce (1877 – 1945), a remarkable American psychic.

Cayce accessed his insights by lying down and easing himself into a still—or sleep—state. After achieving a sleep state, Cayce would answer questions from those physically in the room or via correspondence. This is why he is often referred to as the "Sleeping Prophet." Cayce predicted that the Earth would experience a pole shift event during our times.

To see if this premise can hold water, we begin by looking back through recorded history for evidence of a past pole shift event during Noah's flood. That very evidence is found in *The Kolbrin Bible*.

The Deluge Story of Sisuda and Hanok

Illustration 4: Edgar Cayce

Comparable to the *Torah* story of Noah's flood in Genesis, the deluge story of Sisuda and Hanok is found in the *The Kolbrin Bible: 21st Century Collector's Edition* (Your Own World Books 2nd Edition — May 2006).

In this story, King Sisuda learns that a global cataclysm is imminent and reaches out to his loyal subject Hanok to play a vital role in heading up a government-funded survival ark project.

The Last Pole Shift 15

The full text is contained in the second of the 11 books in *The Kolbrin Bible* and was recorded by Egyptian scribes and academics in the years following the Hebrew exodus. What follows is an abridged version with commentary:

> ***The Kolbrin Bible: 21st Century Master Edition***
> **Book of Gleanings, Chapter 4 – The Deluge**
>
> **GLN:4:16** Then, the wise men... said to Sisuda, the King, "Behold, the years are shortened and the hour of trial draws nigh." ... Then, the king sent for Hanok, son of Hogaretur ... saying, "...the hour of doom is at hand; neither gold nor treasure can buy a reprieve."

King Sisuda tells his loyal subject Hanok that a global catastrophe is coming.

> **GLN:4:17** Then Hanok came into the cities... The governors said, "Go down to the sea, and build your ship there..."

Hanok is told by the governors to build his ark by a sea. In a material sense, this is the smart thing to do. If the catastrophe does not come to pass, you can refit the ark for a commercial application, such as a casino and hotel, and make your fortune. It makes perfect sense —during a period of quiescence.

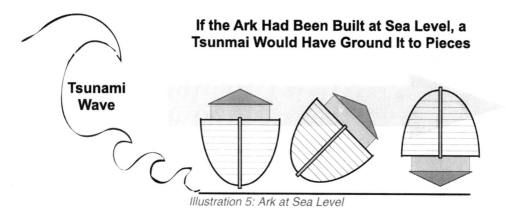

Illustration 5: Ark at Sea Level

However, during a catastrophic event such as a tsunami, the wave will roll over all the vessels in the harbor and grind them to pieces, like hydraulic sandpaper.

> **GLN:4:17** ...But Hanok answered, "It has been told to me in a dream that the ship should be built against the mountains, and the sea will come up to me." When he had gone away, they declared him mad... but... did not hinder him... Therefore, a great ship was laid down under the leadership of Hanok... for Sisuda... from whose treasury came payment for the... vessel.

Hanok knows that he must build his ark in a place ideally suited as an ark launch site during a mega tsunami. He understands the vital need to locate his launch site in a place that will give him the right amount of distance, elevation, and terrain.

These are the three factors that will determine the amount of energy that the oncoming tsunami will expend before it reaches the ark.

Therefore, the ideal place for a launch site is where the tsunami will have expended nearly all its energy by the time it reaches the ark. This way, it first comes into contact with the ark's keel and with just enough energy to float the vessel.

> **GLN:4:18** It was built... close by the river of gold... The length ...was three hundred cubits... its breadth was fifty cubits, and it was finished off above by one cubit. It had three storeys, which were built without a break.

Knowing he would need to launch his ark into the retreating wave of the tsunami, Hanok located his launch site upriver from the coast.

In essence, the Gold River valley would then serve as a natural slalom for the ark's rush seaward with the retreating tsunami—somewhat like a kayaker navigating a white-water slalom.

> **GLN:4:20** Into the great ship they carried the seed of all living things...

Today we have a perfect equivalent. Built into a mountainside in the northernmost inhabited spot on the planet is the Svalbard Global Seed Vault in the forbidding Svalbard archipelago north of the *Arctic Circle*.

The first stone was laid in 2006 for this permafrost fortress, also known as the "doomsday vault." Costing some $6 million, the facility can protect up to 4.5 million seed varieties.

Like a modern-day Sisuda, the key financier of this fortress is Microsoft founder Bill Gates. Others include the Rockefeller Foundation, Monsanto Corporation, Syngenta Foundation, and the Government of Norway, among others.

Given that the current era of petro-dollars will perish in a global catastrophe, the next commodity basis will become the genetic stuff of our God-given food crops—a logical long-term investment for future dominance.

This is because, after the dust settles, those who do survive and want access to the seed bank will have to mortgage their countries to the consortium controlling the doomsday vault. Or, they can choose to starve.

> **GLN:4:22** On the appointed day, they who were to go with the great ship departed from their homes and the encampment. They kissed the stones and embraced the trees, and they gathered up handfuls of the Earth, for all this, they would see no more...

Would people who harbor any reservations, or who cling to hope against hope, kiss stones? Not if they fully understood what was about to happen.

What this account shows is that all who were chosen to survive in the ark had access to the inner-circle insights of Sisuda and Hanok. So powerful were they that they washed away all doubts and false hopes. What a heavy burden it must have been.

> **GLN:4:24** Then, with the dawning... There, riding on a great black rolling cloud, came the Destroyer... she... covered the whole sky above and the meeting place of Earth and Heaven could no longer be seen. In the evening, the places of the stars were changed, they rolled across the sky to new stations; then, the floodwaters came.

The term "Destroyer" is used by both the ancient Egyptians and the Hebrews to describe an object known to the ancient Sumerians as Nibiru, according to Zecharia Sitchin.

According to *The Kolbrin Bible*, previous flybys of this object through the core of our solar system caused the sinking of Atlantis and the Ten Plagues of Exodus.

Today, Nibiru is also referred to as Planet X, Hercolubus, Wormwood, Red Comet, and other names.

> **GLN:4:25** The floodgates of Heaven were opened and the foundations of Earth were broken apart. The surrounding waters poured over the land and broke upon the mountains...

This description goes beyond heavy rains and earthquakes. What this passage actually describes is more likely to be hypercane (massive hurricane) storm surges or mega tsunamis caused by massive landslides and mega earthquakes.

Greater than the power unleashed by the 2004 Sumatra earthquake and Indian Ocean tsunami, as well as Hurricane Katrinain 2005, these regional events will not be separated by years, but by days, hours, or minutes.

> **GLN:4:26** The pillars of Heaven were broken and fell down to Earth. The skyvault was rent and broken; the whole of creation was in chaos. The stars in the Heavens were loosened from their places, so they dashed about in confusion. There was a revolt on high; a new ruler appeared there and swept across the sky in majesty.

This passage clearly describes a pole shift event triggered by a flyby of the object known as the Destroyer. In this, we can we see that the ancient view of the skies changed dramatically, so quickly that it would certainly have appeared as chaos above.

This also raises another possibility—that the rotation of the Earth may have changed as well because this would account for the sun appearing as a new sky ruler due to a different arc across the sky. If so, we must wonder if Earth has always had its current 24-hour day.

> **GLN:4:28** The swelling waters swept up to the mountain tops and filled the valleys. They did not rise like water poured into a bowl, but came in great surging torrents; but when the tumult quietened and the waters became still, they stood no more than three cubits above the Earth.

Also keep in mind that perception is vital. For example, the slow-moving pace of the Pakistan flood of 2010 was used to rationalize the fact that Americans had been 40 times more generous with that year's Haitian earthquake victims than with the Pakistani flood victims.

The author of this passage in *The Kolbrin Bible* is keen to emphasize that this event was more than a flood. This was so future generations would not view it in terms of what they would be most likely to see during a period of relative quiescence.

He then goes on to describe a mega tsunami which will come in a torrent of approximately seven waves. What is also interesting to note here is that this event triggered a substantial rise in sea levels.

Depending on how one calculates a cubit, the immediate rise in sea level due to the event was between 4.5 ft. (1.37 m) and 6.75 ft. (2.05 m). This was likely due to the convergence of two factors.

First was the loss of major ice sheets like we're now seeing in Antarctica and Greenland. However, the other factor could be just as powerful, if not more, as land masses will rise and fall during a pole shift event.

The reported sea-level rise of as much as 6.75 ft. (2.05 m) in such a short period of time does suggest a catastrophic one-two punch, replete with severe, ice sheet loss and both rising and sinking continents.

> **GLN:4:28** ...The Destroyer passed away into the fastness of Heaven, and the great flood remained seven days, diminishing day by day as the waters drained away to their places.

This passage accurately pinpoints the onset of these events at or near the time the Destroyer approaches its point of perihelion (closest distance to the sun) during its orbit through the core of our system.

What is important to remember with a long-period, elliptical orbit is that as the object draws close to the sun it speeds up. Consequently, its passage through the night sky at that point is faster than it is during the long voyages to and from the core of our system.

To help you visualize this, imagine yourself at a local amusement park and you've just handed the ride attendant a ticket to ride The Whip.

A long-standing popular ride, The Whip moves you around much like an object in a long-period orbit. You coast toward the pivot point as your anticipation builds. Then, you're whipped about and the next thing you know, you're coasting in the opposite direction.

> **GLN:4:28** ...Then, the waters spread out calmly and the great ship drifted amid a brown scum and debris of all kinds.

In this passage, we can see the extent of the topsoil erosion and violent deforestation.

The mention of brown scum is also quite familiar given the catastrophic and far-reaching 2010 BP oil spill disaster in the Gulf of Mexico.

This indicates that massive subsea earthquakes along with rising and falling continents will unleash vast amounts of undersea oil.

> **GLN:4:29** After many days the great ship came to rest... in The Land of God.

As with the story of Noah's flood in Genesis, the ark of Sisuda and Hanok drifts until it runs aground. This indicates that the event was not like a flood, where the waters eventually return to a former state. Rather, the ark had to drift to high ground far from its initial launch site.

This clearly demonstrates a massive, permanent, and profound change in the lithosphere (the crust and a portion of the upper mantle) of the planet.

What Triggered the Deluge?

The Destroyer, an object in orbit around our Sun, is the cause described in the deluge story of Sisuda and Hanok.

Aside from noting that this object appeared coincidental to the deluge event, the accounts fail to explain how the Destroyer actually wrought such havoc. This omission is not a legitimate reason to dismiss this wisdom text, however; it simply fails to explain the event with 21st-century clarity. We're equally vulnerable to such criticism.

Let's assume we're driving down a lonely road and our car motor dies. We steer off to the side of the road as the motor sputters.

Although we may notice the flashing warning lights on the dash, most of us are in the dark as to why the motor dies. Even if we're mechanically inclined, this holds true if we're without our modern testing tools.

Hoping for luck, we lift the hood to see if the failure is obvious and simple to fix. In most cases, we're not so lucky, so Plan B kicks in and we try to call for help or flag down a passing motorist.

Eventually, some people show up with the right tools and they know what needs to be done. They explain the problem to us well enough that we know what we're about to pay to fix.

The points illustrated by this motoring analogy are that we need to:

- Know what to look for.
- Have the right tools to interpret the signs.
- Formulate a plan of action.
- Act on that plan.

In this case, there are clear signs that Planet X has been associated with a previous deluge and that it will trigger another pole shift with its next flyby through the core of our solar system.

3

The Trigger Event

Simply put, the trigger event for the pole shift is described in a crop circle that appeared in two parts, on the 15th and 22nd of July 2008, in Avebury Manor near Avebury, Wiltshire, UK.

Over the last decade, numerous crop circles have appeared worldwide, but principally in England. A few have appeared with urgent warnings or predictions of things to come. Like the ancient prophets and seers, these predictions come with harbingers. Such is the case with the Avebury 2008 formation,

Unlike most formations which catch the eye with exquisite geometry, this two-part formation is truly a first-of-a-kind communication event.

It gives us a consistent and sophisticated Nibiru / Planet X prediction that describes where this object will be seen, from where, and when. When that day occurs, the unprepared will bear the brunt of the suffering. During

Illustration 6: Avebury - July 22, 2008

this period, horrific solar storms will scorch the Earth, and following on the heels of that will be a pummeling of asteroids—and not just once.

Altogether, these catastrophes will set in motion the Earth processes that will result in a pole shift of the planet, as predicated by Edgar Cayce.

The disinformation war waged against this formation is considerable. Yet, when one views this critical formation without the simplistic distractions of disinformation, the messages snap together cleanly and comprehensively.

The key is an objective perspective, free of economic, political, and theological agendas. Simply view the formation with the eyes of a child. Read it just as it was created. When you do, you come to realize the intent of the crop circle makers—that being to help us, perhaps, once again.

Crop Circles, Deep Time History, and Global Catastrophes

From a familiar Biblical perspective, global catastrophes occur within a relatively short time, such as the 40 days and nights of rain mentioned in the Genesis account of Noah and the flood.

In terms of creating an enduring allegory, simplistic concepts such as this are vital to helping future generations. They may never witness such an event, but they can embrace the story and, most important, pass it on.

However, what the deep time history of our planet shows is catastrophic events that are lengthy and far more complex in terms of causes, interactions, and consequences. In essence, we can use a jigsaw puzzle to explain.

Allegorical Biblical and folklore accounts show us the big picture on the top of the puzzle box so that we can take it or leave it as we choose. Those interested can dump all the pieces onto the coffee table and start sorting out the corners and edges of the puzzle.

Others, with their own interests and agendas, may glance at the puzzle, but if it does not serve their pursuits, they'll leave it sitting on the shelf. The result is that most people are willing to dismiss crop circles with nothing more than a glance. Mainstream science teaches them to respond this way.

Like medieval grand inquisitors, dismissive scientists often label the study of crop circles as "bad science" and other such diminishing terms.

Inevitably, the mode of attack follows a predictable pattern, where accusations are never substantiated beyond opinion. Rather, the public is told that scientists know better and that there is nothing to learn. For those seeking a simplistic answer to quell their interest, this works remarkably well.

A good example is the *National Geographic* (NGC) cable television channel. A few years ago, NGC sent video production teams to England to document crop circle phenomena.

Well-known and highly regarded crop circle researchers greeted them with open arms. They were eager to support what they were told would be an even-handed and intelligent examination of crop circles. They were duped.

Time and again, the programs that did air revealed the agenda-drive, medieval bias of NGC and little else. The hard science provided by the legitimate researchers interviewed was mostly left on the cutting room floor, save for the occasional slanted sound bite. The real focus was a propaganda agenda bent on suppressing an educated awareness of the phenomena.

The result is that most genuine crop circle researchers now avoid any interview requests from NGC. As the old saying goes, "fool me once, shame on you; fool me twice, shame on me." Their wariness also extends beyond NGC to the hit-and-run debunkings and disinformation-for-profit tactics of the entertainment industry. A case in point is the Disney/Touchstone science fiction thriller *Signs* (2002).

Written, produced, and directed by M. Night Shyamalan and starring Mel Gibson, it was a 1950s-style science fiction spoof of crop circles. The entertainment goal was to scare audiences, and Shyamalan rehashed the hackneyed cinematic formula of aliens invading Earth to harvest humans as food.

Likewise, the July 2008 Avebury two-part formation has been broadsided with hit-and-run debunkings and disinformation. A case in point is Freddy Silva, who appeared on "Cut to the Chase," the author's Internet radio show, in 2004.

Cut to the Chase #17
Host: Marshall Masters
05-Aug-2004 [0:57:55]

Crop Circle Researcher Freddy Silva Shares Bold Insights on the 2001 Arecibo Reply and 2002 E.T. Formations

In 2009, this author contacted Silva regarding the July 2008 Avebury two-part formation. What follows is the actual email conversation that occurred over a span of two days.

Illustration 7: Cut to the Chase, Show #17

Wed, Oct 7, 2009
FR: MARSHALL MASTERS
TO: Freddy Silva

Hi Freddy:

Been a long time since you've been on Cut to the Chase. Would like to have you back to discuss 2 things. Your video: ORBS, CROP CIRCLES and SACRED SITES: The Manifestation Of Soul? And, to discuss the Planet X Avebury 2008 formation. Are you available later this month, or next?

Thanks, Marshall

Wed, Oct 7, 2009
FR: FREDDY SILVA
TO: Marshall Masters

Sure.

I'm off on a short tour and will return next week, so, send me some dates. 2008 at Avebury: fake.

Wed, Oct 7, 2009
FR: MARSHALL MASTERS
TO: Freddy Silva

Why a fake and by whom?

Wed, Oct 7, 2009
FR: FREDDY SILVA
TO: Marshall Masters

See my 2008 report, under HISTORY.

Off to pack...

Wed, Oct 7, 2009
FR: MARSHALL MASTERS
TO: Freddy Silva

I could not find a history link on your new site. Could you please provide a URL?

Thanks, Marshall

Wed, Oct 8, 2009
FR: FREDDY SILVA
TO: Marshall Masters

http://www.cropcirclesecrets.org/history.html

Wed, Oct 8, 2009
FR: MARSHALL MASTERS
TO: Freddy Silva

I checked out the page and it does not say anything specifically about Avebury. Please provide specifics.

Thanks, Marshall

For the record, Freddy Silva has never responded to the author's request for specifics, nor has he published his findings as of the publication of this work. Rather, once his hit-and-run debunking of the formation was called into question, he disappeared.

It is interesting to note that inquiries to other legitimate crop circle researchers about Silva's findings yielded a consistent result. Silva never visited the site.

However, those researchers who walked and/or flew over the formation all reported that their findings clearly showed it to be a genuine formation. In fact, that year it was one of the most widely discussed formations by the UK crop circle research community and elsewhere.

Two of these researchers featured their own flyover videography of the formation in their award-winning video documentaries. They were Patty Greer and Suzanne Taylor, and both appeared on the author's "Cut to the Chase" program in 2009 to discuss their videos and their own impressions of the Avebury formation.

Cut to the Chase #106
Host: Marshall Masters
11-June-2009 [1:16:17]

The 2008 Avebury Planet X and 2012 Crop Circle Ground Report — Award Winning Documentary Producer Patty Greer

PattyGreer.com

Illustration 8: Cut to the Chase, Show #106

Cut to the Chase #116
Host: Marshall Masters
19-January-2010 [01:10:32]

The Who, What, and Why of the Crop Circle Makers — Documentary Producer Suzanne Taylor

WhatOnEarthTheMovie.com

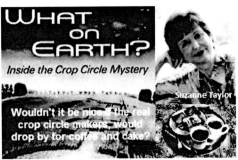

Illustration 9: Cut to the Chase, Show #116

This dichotomy between Freddy Silva, the debunker who neither visited the formation nor published any findings, and the professional and considerable efforts of Patty Greer and Suzanne Taylor is stark.

The only thing Mr. Silva offers is an uneducated and expedient way to dismiss the formation and, regrettably, there are plenty of takers. However, those who have connected with this formation in a genuine way see something much greater than Mr. Silva's expediency—a SPOT of immediate interest.

Single Point of Truth (SPOT)

There are times of synchronicity where the universe throws us a single point of truth (SPOT). Like picking at a loose thread, a SPOT can unravel the most complexly rolled ball of string—that is, with patience and determination. Such was the case when this author started tracking the Avebury formation after the first part formed on July 15, 2008.

The second part formed as a partial overlay of the first on July 23, 2008. A day later, the UK-based Crop Circle Connector (www.cropcircleconnector.com) website, run by Mark Fussell and Stuart Dike, posted an update on the formation.

Less than an hour after that update was posted, this author received an alert from a trusted source with a simple message—Get to the page before someone gets to them—and the URL for the update.

Shortly after this author took a full screen capture of the update page, the Crop Circle Connector web server began reporting error 404 "Page not found" for the update URL. An email inquiry as to the page's disappearance was immediately made to Fussell and Dike, but neither responded.

The consequence was stunning. Here was a significant event in crop circle research and the entire update web page, along with all of the graphics, had simply vanished.

It was as though they'd never been there at all. Sensing foul play, this author published an expose on July 26, 2008.

> **YOWUSA.COM, 26-July-08**
> **Are Governments Suppressing the 23-July-08 Planet X / Nibiru Crop Circle in Avebury?**
>
> The Crop Circle Connector website has long been a leading UK crop circle resource, but a recent, mysterious retraction of what could be a groundbreaking formation depicting the orbit of Planet X / Nibiru now begs the question: Is this the result of government suppression? Are we being prevented from learning about the Planet X / Nibiru Crop Circle that appeared in two parts in Avebury Manor, near Avebury, Wiltshire, UK on 15-July-2008 and 23-July-2008?

Without any fanfare, notice, or explanation, the update page was restored by the Crop Circle Connector website on August 5, 2008. It was reposted with additional images and elaborate interpretations that often did more to confuse readers than to help them.

Did Fussell and Dike intentionally engineer these odd circumstances? This author unequivocally says no. These men are respected and admired by crop circle researchers for their enduring commitment and professionalism, and rightfully so. Furthermore, this author encourages you to visit their site and to generously support their efforts.

That being said, at the end of the day, they're just two guys in the UK running a website. As such, they are easy pickings for someone with deep pockets, extensive resources, and the will to use them.

The Struggle with Planted Disinformation

In preparing the findings on this two-part formation you are about to read, it is important to note that the author, with the help of several other independent researchers, has studied this formation for more than two years.

One pattern continually emerged each time the research group attempted to reconcile its preliminary findings with those published in the commentary on the Crop Circle Connector site. The result was always the same: confusion.

It was a perpetual Tower of Babel scenario. For example, one field diagram of the complete formation provided put both Earth and Mars in the same orbital path. Yet, another version of the graphic placed both planets in their correct orbits. Another difficulty with the analysis of this two-part formation was that all the aerial photos were taken by observers in low-flying ultra-light aircraft.

Ultra-light pilots are wary of prolonged crop circle over-flights and with good reason. Genuine formations cause electromagnetic interference that can disrupt their flight instruments and radios. Such was the case with the Avebury 2008 formation, and so the perspectives of the available aerial images lack a truly perpendicular, look-down view. However, the commentary was the confusing part.

While a good deal was highly useful, much of the commentary injected distractions and insoluble puzzles. Consequently, one can read the page dozens of times and never see the dots connect, even though a much larger story is in plain view.

The question is: How does one find it? The answer is simple. Distraction, whether intentional or not, is still distraction. It's noise.

For example, in the book *Planet X Forecast and 2012 Survival Guide* this author stated:

Planet X Forecast and 2012 Survival Guide
Your Own World Books (2007)

Illustration 10: Planet X Forecast

...there is a load of great information on TV, especially stories about weather and habitat change. To get the most out of them, watch them like a deaf person. Turn off the sound and turn on the closed captioning. Then play something soothing like Debussy's Claire de Lune on the boom box, and take written notes. You'll be amazed at what you really can notice when you're not listening to theme music that's designed to pump up your heart rate.

> Do not worry about missing anything. What you
> really want are the repeating visuals; the video
> segments and graphics that are continuously
> repeated with something new occasionally thrown
> in. Over time, you'll begin to notice different
> things, especially when you surf the news
> channels.

After two years of studying the formation, the possibility of planted disinformation finally had to be acknowledged. Why? It was the only conclusion consistent with Occam's hammer: The simplest answer is usually the right one.

Therefore, the solution was to treat the Avebury commentary no differently than a TV program. Turn off the noise and focus on the images.

With this change of tactic, the result was immediate and stunning. The genius of the crop circle makers became self-evident, along with the powerful and comprehensive message they intended for us to receive. Now we have! What follows are the findings of that analysis.

Avebury Part 1 of 2

There are several impressive aspects to this two-part formation, and the most remarkable is the manner in which the first part is later overlaid by the second.

July 15, 2008 – The First Part Appears

The first part of this Nibiru / Planet X formation at Avebury Manor near Avebury, Wiltshire, UK.

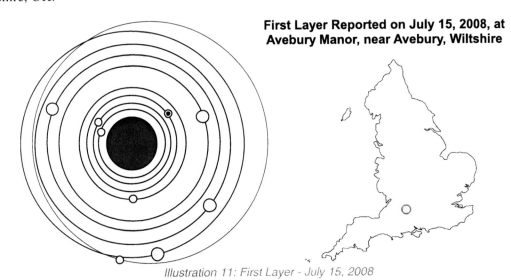

Illustration 11: First Layer - July 15, 2008

Though crop circles have formed worldwide for centuries, one could say that England is to crop circles what the Napa Valley of California is to winemaking—a perfect place to express a grand idea.

This is why every summer researchers look toward England with great anticipation, for in these fertile fields not far from Stonehenge, many of the greatest formations will appear.

An Asymmetrical Formation

What is immediately obvious in the first part of the Avebury formation is its asymmetrical design; most formations are symmetrical. However, both types are equally beautiful and informative.

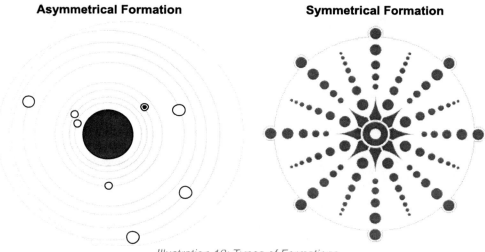

Illustration 12: Types of Formations

Symmetrical formations speak to us with what is commonly known as sacred geometry. These formations favor the left side of our brains.

In contrast, asymmetrical formations speak with symbology. They

require a balanced view, using both sides of the brain. Both types pique our curiosity by tapping into the very consciousness of our species—if we are willing to accept the call.

It is also worth noting that symmetrical formations principally convey a single message, but asymmetrical formations such as Avebury can simultaneously convey multiple messages.

Winter Solstice Time Stamp

From the outset, one message in the first Avebury formation immediately catches the observer's eye—the celestial alignment of the major planets: Mercury, Venus, Earth, Mars, Jupiter, Saturn, Uranus, and Neptune.

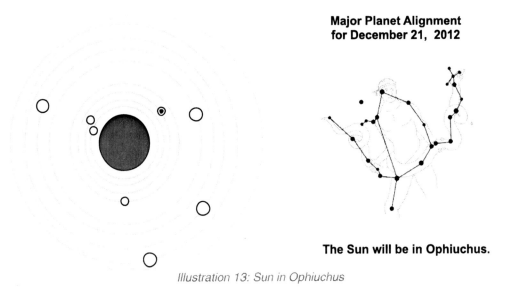

Illustration 13: Sun in Ophiuchus

It is a clear and unequivocal time stamp: the winter solstice on December 21, 2012. It is also important to note that on this day, the Sun will be in Ophiuchus. What is interesting about Ophiuchus is that it is a recurring element in Planet X research. A good example is The History Channel's documentary, "Lost Book of Nostradamus" (2007).

The program explores newly revealed writings from the 16th-century French astrologist Nostradamus and attempts to explain a series of illustrations with predictions for today. In these illustrations, Ophiuchus plays a key role.

It is interesting to note that the last illustration described in the documentary clearly demonstrates that Nostradamus is showing us that our view of the sky will change during this event, which is clearly a consequence of a pole shift event.

Avebury Part 2 of 2

When the second part of the Avebury formation appeared days later, the resulting overlay of parts one and two created a breathtaking formation—a first of its kind.

Like the Dresden Codex, which revealed the 2012 prophecy of the Mayan calendar to Western cultures, this completed formation in Avebury is also encoded with its own brand of interconnected messages about this same moment in time.

July 22, 2008 – The Second Part Appears

The second part of this formation was first reported July 22, 2008. Unlike the first part, which was a mix of symmetrical and asymmetrical aspects, this one was mostly asymmetrical —that is, except for one mysterious, big empty ring.

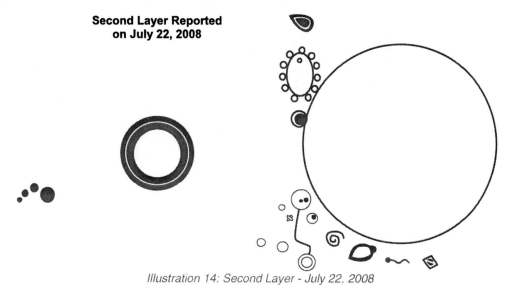

Illustration 14: Second Layer - July 22, 2008

The big empty ring immediately stands out in contrast to the other glyphs within the second part, all of which offer symbolic messaging.

Nonetheless, in terms of messaging, the second part of the formation is far more complex than the first, which principally served as a predication date stamp for December 21, 2012.

The Complete Formation

One of the most amazing aspects of the second part of this formation is how the visual representation of the Sun expands from its original size in part one. With the overlaying of part two, it engulfs Mercury and reaches out to the orbit of Venus.

Multi-part formations like Avebury are uncommon, but not until Avebury 2008 had crop circle makers employed a staged, partial overlay to send a complete message.

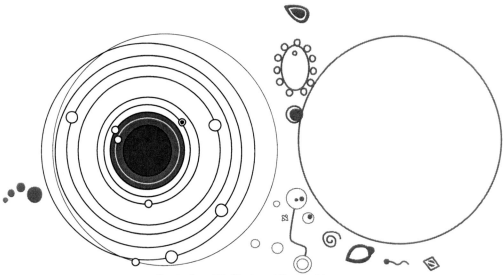
Illustration 15: First and Second Layers

Avebury 2008 was without doubt a first-of-its-kind for two reasons. First, it shows that the Sun is going to become more violent. Of equal importance, it demonstrates the precise, pinpoint control of crop circle maker technology. It is stunning when you consider the size of the entire formation.

Size Does Matter

According to crop circle field researchers who walked and overflew the formation, it was nearly the size of four soccer fields. They also reported that traversing the formation on foot from end to end took approximately one hour.

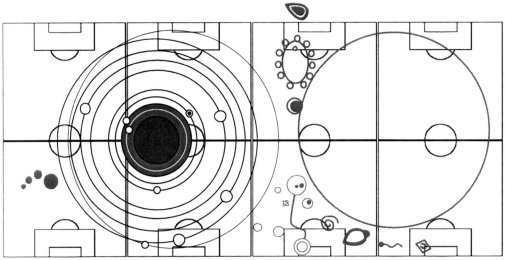
Illustration 16: Size of Formation

The size of the Avebury formation is even more impressive, given that genuine crop circles form in approximately 8 seconds, as documented through actual observation. So, it took less than 20 seconds for the crop circle makers to create Avebury 2008—a two-part, overlaid formation that is nearly the size of four soccer fields and laden with clear, prophetic messages.

Yet this massive formation also has a rather mysterious aspect as well.

The Minestrone

The most mysterious part of the second part of the Avebury formation is nicknamed "The Minestrone" by the author. It is a series of apparently unrelated glyphs wedged between the two largest glyphs of the formation. There are many explanations to connect these glyphs vis-a-vis various celestial alignments, mythology, folklore, and beliefs.

Illustration 17: Harbinger Symbols

While some can explain them in part, none has found an exception-free, unifying theory that connects the dots between these various minestrone glyphs. As with all things, there is a reason behind this as well.

Throughout history, the prophecies of great prophets and seers have always come to us in two parts. The prediction and its harbingers. Both aspects serve the same message, but at two different times.

The same holds true for this formation. These mysterious if not obscure glyphs seem to be tightly wedged into a bit of convenient free space on the greater canvas of the formation. Yet, there is an element of time in the manner of their placement.

There are many harbinger glyphs, each small enough to be easily scattered across the greater formation. However, the circle makers chose the placement so that we would not miss the point—that they are the harbingers of this formation.

In time, their meanings will become as obvious to us as the rest of the formation is today. When they do, we will be able to connect the dots. Then we will know that we have passed a point of no return.

Yet, there is something immediately useful in the placement of these harbinger glyphs and understanding that is like driving a tractor across a farmer's field.

Tramlines in the Formation

Given that the Avebury formation was nearly four soccer fields in size, it would naturally span across several of the tramlines in the field. Tramlines are the parallel lines in the fields on which farmers drive their tractors. They also give farmers a way to fertilize and spray without damaging surrounding plants.

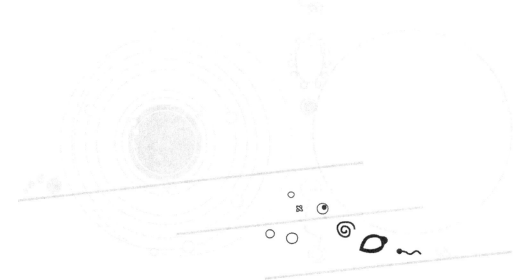

Illustration 18: Tram Lines

For those who want to enter a crop circle formation on foot, tramlines offer easy ingress and egress. Best of all, there is no unnecessary trampling of plants, which are the property of the farmer.

Hoaxers use the same tramlines for ingress and egress as well, which is important to keep in mind. This is because hoaxers know they can make a more convincing fake if they work from the tramlines.

Consequently, tramlines have become one of the most useful tools for field researchers when authenticating a formation. They look for signs of unintended ingress and egress, and here is where some of the minestrone glyphs hit real pay dirt.

Depending on how they are counted, there are at least seven or more glyphs between the tramlines. They are sufficiently distant from the tramlines and each other so that a hoaxer could not leap from one to the next.

These between-the-tramline glyphs were quickly noted by researchers like Patty Greer. According to her, there were no signs of ingress or egress to and between the tramlines and the glyphs. In other words, it would have been impossible for a hoaxer to create these between-the-tramline glyphs with his or her feet on the ground.

However, there is one glyph in the second part of the formation that is mystifying, depending on your point of view. It is the large, empty, oval ring.

The Avebury Binocular

When folks first look at this completed formation, it is like looking at the front lens of a pair of binoculars. One is very busy and informative while the other is totally blank. Why let that much prime real estate go to waste?

Illustration 19: Ecliptic Ring

Also, the makers of this crops circle went to great effort to compact the minestrone in what was clearly a small wedge of the overall formation canvas. All of those glyphs would have fit nicely inside the big empty ring.

Or, could this be another less-is-more situation?

After all, one thing crop circle makers have always shown is that there is a purpose to everything they do. This is why you never see an unfinished formation, even when it must be staged in multiple parts as at Avebury.

Ergo, the appropriate question is: What kind of purpose could be so important that it requires this much prime real estate to convey?

To answer that question, you must first stand your ground and see what you intend to see. When it comes to observing the sky, the ecliptic is, in a manner of speaking, your farthest horizon.

Understanding this simple concept is essential to your being able to visualize how Planet X moves.

The Ecliptic

In terms of observing Planet X, a basic understanding of the ecliptic is essential as it helps us to visualize the object's path through the inner core of our solar system.

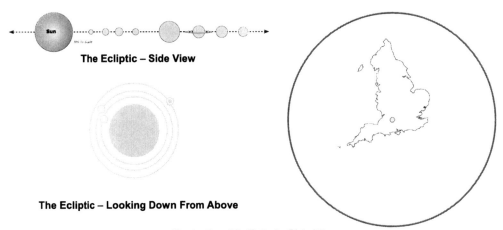

Illustration 20: Ecliptic Side View

To imagine the ecliptic of our solar system, start at a point in the center of the Sun and then expand a concentric ring from there out to all 12 signs of the zodiac.

That is the ecliptic. Consequently, those living above the equator see a very different sky than those living below the equator.

Ergo, if you are in living in Los Angeles, you cannot see what folks in Sydney are seeing in their sky and vice versa. This is because the curve of the Earth gets in the way. Consequently, before December 21, 2012, folks living down under will see the Destroyer earlier those of us living above the equator.

That being said, it is interesting to note the location of the Vatican Advanced Technology Telescope (VATT) at the Mount Graham International Observatory (MGIO) in southeastern

Arizona. This site is in a prime viewing band for clear skies and is ideally situated for the longest possible view of Planet X as it transits the core of our solar system.

For those of us who cannot afford to build an observatory in Arizona, we just need to know where to be on December 21, 2012, so that we can see the harbingers and predictions in this formation fulfilled. We just need to know when to be there and where to look—and this information is provided by the crop circle makers.

Where?

In terms of the first part of the formation, it shows us a celestial alignment of the major planets on December 21, 2012. As we have the when, the inquiry must now focus on the where.

Given that this is an astronomical observation, "where" is a two-part question. First, what part of the sky will you be looking at? Second, from where on Earth will you be doing your looking?

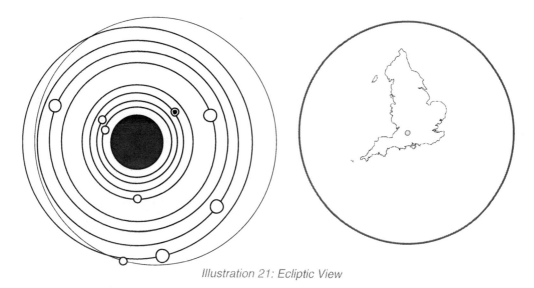

Illustration 21: Ecliptic View

This is why the big empty ring left in the second part of the formation quickly gives us a spot-on answer to the question of where to stand when gazing into the heavens for the celestial alignments depicted in the formation. Herein is the simple genius in this glyph design.

At first glance, it may seem natural to think that an x-marks-the-spot symbol, such as a big X glyph carved inside the ring, would send a clear message. However, in practice that's not likely, as it would make it too easy to waste time rummaging through rabbit-hole assumptions.

Conversely, an empty ring sends a clear and unequivocal x-marks-the-spot message. This is because there can only be one thing inside the ring—good old, jolly old England itself or, more specifically, that little spot of land in Avebury within the inner confines of the empty ring. This is the "where" stamp of the second part of the formation that connects it to its twin.

Ergo, the complete when and where message is: Stand on this spot on the waning crescent phase of the moon for the month of December 2012. As you do, hold your hand straight out with the finger pointed at the center of the moon.

Then track your arm horizontally 45° degrees to the right of the moon, at which point your finger will be pointing at the area of the sky where Planet X is to be found.

The only remaining questions at this point are where to look and for what. Sadly, the answers to those two remaining questions will be the causalities of our greatest woes.

Path of the Destroyer

The first thing that becomes obvious in the second part of the formation is the use of a recurring metaphor to indicate an incoming object in a long-period elliptical orbit. It also shows us its relative point of entry into the core of our solar system.

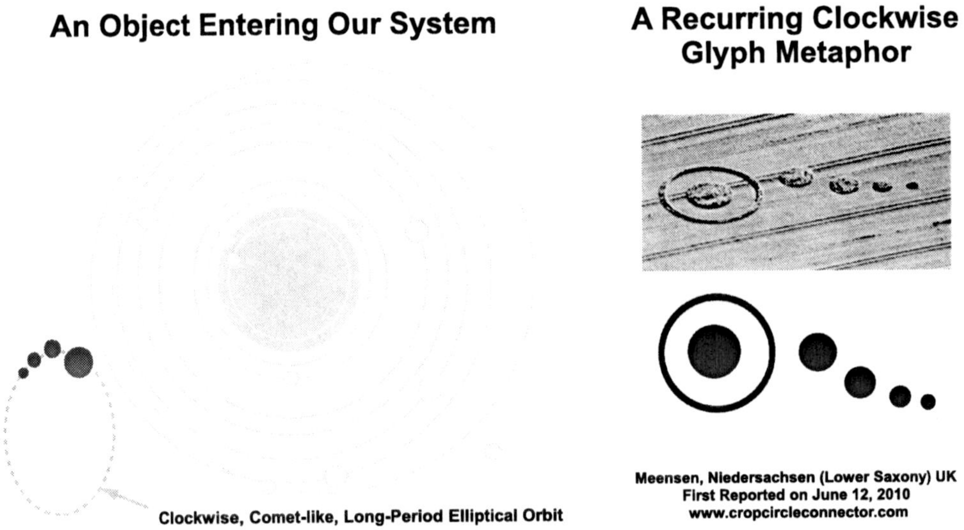

An Object Entering Our System

Clockwise, Comet-like, Long-Period Elliptical Orbit

A Recurring Clockwise Glyph Metaphor

Meensen, Niedersachsen (Lower Saxony) UK
First Reported on June 12, 2010
www.cropcircleconnector.com

Illustration 22: Planet X Inbound

This same visual metaphor is also used in a small formation that was first reported on June 12, 2010, in Meensen, Niedersachsen (Lower Saxony), UK. Like the 2008 Avebury formation, it too shows an object in a long-period elliptical orbit approaching another body in space.

In both cases, the metaphor shows us the object's inclination to the ecliptic. This is important because it helps us understand why this object seemingly will be able to sneak up on us.

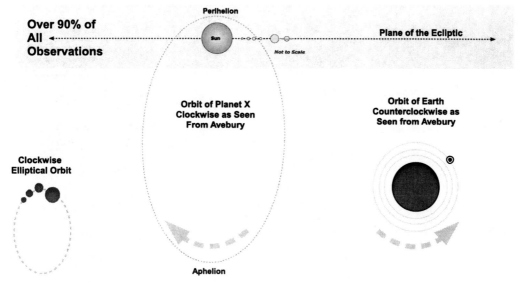

Illustration 23: Planet X Inbound

The term "Planet X" is a long-standing generic term used by astronomers and mathematicians. It describes how an unseen object first reveals itself by the manner in which it interacts with other well-known and observed objects.

This is why this author prefers this term to describe this object, which is also known and discussed by various other names, including Wormwood, Nibiru, Destroyer, Frightener, and Red Comet, among others.

It is further interesting to note that the crop circle makers are aware of our present understanding of ancient symbology for this object, as evidenced by another Avebury formation first reported August 16, 2010.

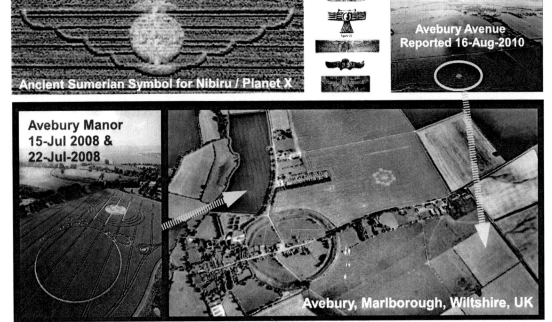

Illustration 24: Symbol for Planet X

Etched in the maze and cataloged as "Avebury Avenue, nr Avebury, Wiltshire. Reported 16th August," this formation represents the ancient Sumerian symbol for Nibiru / Planet X. Two aspects of this new formation are quite remarkable. First, it formed within one kilometer of the Avebury 2008 formation and, second, both formations are about the same incoming object.

While the distance between the 2008 and 2010 Avebury formations is a relatively light and speedy stretch of the legs, so to speak, the same cannot be said for this object.

It is vital to note that the crop circle makers are clearly telling us it is in a long-period orbit and currently approaching the core of our solar system from beneath us. This helps to explain why this object can be so difficult to observe, especially for those who assume that astronomers view every part of the sky with equal regard.

A Candlelight Conundrum

The truth of the matter is that 90 percent of all observations occur within a few degrees above and below the ecliptic. The other 10 percent of observations are divided between the southern and northern hemispheres, with observations in the northern hemisphere garnering the lion's share of that small percentage.

This object is approaching from a part of the sky that receives less than 5 percent of our overall attention. Furthermore, its orbital path takes it through a very dense area of the sky.

Consequently, assuming we can see it in the range of visible light at this time would be much like spotting one candle against the visual backdrop of a huge public candlelight ceremony.

Distance also plays a key role. The object's distance to the Sun determines what we can see with our own eyes and with amateur telescopes as well. While amateur astronomers do a marvelous job of discovering comets, they only do so once comets' trajectories puts them inside the orbit of Jupiter and that is when they light up.

At present, the best estimates for this object place it at twice the distance from our Sun as Jupiter, so it will be a while before seeing it becomes possible.

While amateur astronomers in the southernmost locations of the globe will be able to observe the object before it becomes visible to the naked eye, there will be heavy-handed suppression of these reports. A clear example of this was the media coverage of the 2010 BP oil well catastrophe in the Gulf of Mexico.

Those who review CNN footage can hear its reporters consistently report that they'd been denied access by BP security personnel—with the support of government enforcement officers.

Other reports of independent documentary filmmaker arrests and access denials to key areas were also frequently reported. The result was a real-time whitewashing of any viable independent reporting so as to suppress any that would be contrary to BP and government interests.

It would be naïve to assume that this powerful real-time whitewashing mechanism would not be similarly used to suppress independent amateur astronomer observations. Funded astronomers will feel the political heat as well. Consequently, this object will not be formally announced until it is barely visible to the naked eye. The wise will expect that.

With this in mind, the harbinger glyphs described by this author as the minestrone in this formation played an immensely valuable role. They allow us to transcend the natural barriers of astronomical mechanics and the manmade barriers of governments and special interests. So, remember them and be vigilant. When they become obvious, we have crossed a point of no return.

After the cataclysm is spent, it will take time for the skies to clear, but they will. Then those who survived will begin to search the skies for old friends, but in new locales. They will find one old friend, Pluto, and it will be a bittersweet discovery.

Pluto's Fate

Pluto is a truly incredible planet, and many of us have an inexplicable affection for this cold and distant rock smaller than our own moon. People took to the streets and protested when Pluto was demoted from a major planet to a dwarf planet.

Perhaps inexplicable affection is the result of Pluto being the runt of the litter. Isn't it always true that the youngest one in the family seems to get special affection? With this in mind, it is sad to see what the Avebury formation predicts for Pluto.

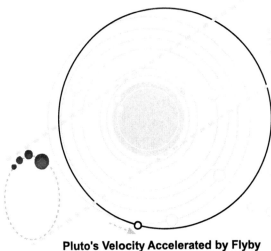

Illustration 25: Pluto Perturbed

Crop circle researchers quickly picked up on the placement of the Pluto symbol relative to its future orbital position on December 21, 2012. Although Pluto's trajectory is not changed, its velocity is, and markedly so.

What accounts for this? Pluto's increased velocity is the result of Planet X crossing inside of Pluto's orbit on its way to the center of our solar system.

The eventual end result is going to be that we will lose Pluto. It will either spin out into deep space or go into a death spiral toward the sun.

The Kozai Mechanism

What explains Pluto's change in velocity is something called the Kozai mechanism, a resonance effect first described in 1962 by Japanese astronomer Yoshihide Kozai. He made this discovery while analyzing the orbits of asteroids.

The Trigger Event 45

For the purposes of this discussion, the Kozai mechanism explains how an unstable Planet X can perturb the stable orbits of other bodies in our solar system. One way to visualize this is to imagine a highly polished marble tabletop.

On the table we see a dozen spinning tops such as we enjoyed as children. There they are, spinning happily and neatly in place, each with its own generous boundary on the table.

Then from one corner, a drunken interloper comes spinning in, wobbling and skittering as it jumps about. As it skitters, one stable top after another is destabilized. Some wobble and then flop on their sides. Others skitter drunkenly off the edge and fall away.

In terms of Avebury 2008, the crop circle makers are telling us that Planet X is our interloper and that our old friend Pluto is its victim. Yet there is an even more profound message here: *Time is of the essence for humanity.*

These flybys have occurred in the past and are well documented in folklore and wisdom texts from all four corners of the world.

However, this flyby could be the next to the last. In the time beyond ours, the last flyby will result in the end of all life on this planet. What would be the date? The best estimate so far comes from Nostradamus. The ancient seer put the end of the Earth as late as 3797.

"Hold on a second," some might say. "If this object has an orbital period of 3,600 years, how could it possibly occur again in 3797 when it should be here again in 5612, by the numbers?" At face value, the question is legitimate given a difference of 1,815 years.

The Math Just Doesn't Add Up!

The reason why the math does not add up is that it is the wrong kind of math. Simply adding and subtracting numbers is often clumsy and misleading because this approach assumes that a two-dimensional calculation can easily explain a three-dimensional reality. A good example is Comet Hale-Bopp, the Great Comet of 1997.

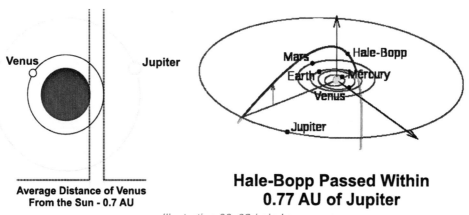

Illustration 26: 03-hale-bopp

An AU (astronomical unit) is the mean distance between the Earth and the Sun (Earth is 1 AU from the Sun). Using the same measure, the mean distance between the Venus and the Sun is 0.7 AU.

In 1996, Comet Hale-Bopp passed within 0.77 AU of Jupiter. In terms of shear mass, Jupiter is 2.5 times the mass of all the other planets in our solar system.

When Hale-Bopp came within a distance of Jupiter roughly equivalent to that between Venus and the Sun, it is believed that the gas giant's tidal gravitational forces caused as dramatic change in the comet's orbit. The orbital estimates show nearly a 40 percent reduction in the comet's orbital duration as a direct consequence of this 1996 flyby past Jupiter.

	Orbit Estimates for 1996 Jupiter Flyby	
	Pre-1996	**Post-1996**
Perihelion every...	4,200 years	2,533 years
Aphelion distance	525 AU	370 AU

During this coming flyby, the orbit of Planet X could likewise be perturbed through a combination of solar interactions and the tidal gravitational forces of Jupiter. The result could be that the present orbit for Planet X deteriorates and, upon its return, it plows through our solar system just like our spinning tops on the marble table.

In this far future time, Planet X could then destabilize the orbits of other bodies, as with Pluto according to the Avebury formation. In other words, 2012 could very well plant the cosmic seeds of Earth's demise in 3797, with a violent, planetary extinction event.

Therefore, let us not go the way of Pluto because we failed to imagine the unimaginable. Rather, we possess the capacity for free will and this is the most essential meaning of 2012—that we are free to choose how we rise to this challenge and to what end. May we do so wisely. In the meantime, Planet X already has its own path to follow.

The Path of Planet X

When creating illustrations, a common technique employed by desktop design artists is to align objects using one or more snap lines. Though each object may feature a different size and dimension, this simple alignment technique helps to present them as a more eye-pleasing image.

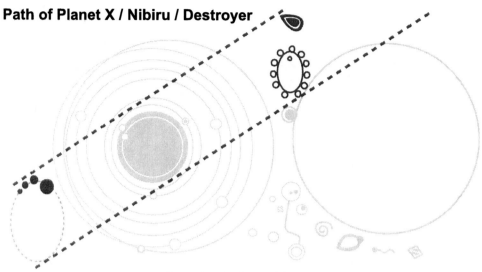

Illustration 27: Path of Planet X

In this respect, it is as though the crop circle makers, like artists, understand how we visually scan images. This is because they clearly used this technique to help draw our attention to the most important glyphs in this formation—the three principal glyphs that speak directly to Planet X.

The one in the lower left-hand side of the illustration shows an elliptical clockwise orbit metaphor. At the upper center is the comet-like Planet X glyph. Below that, another glyph pinpoints where Planet X will be relative to the Sun in December 2012.

The Whip

The glyph that shows us where Planet X will be relative to our Sun in December 2012 looks somewhat like a conference table with round chairs distributed around it. Each of these represents various points of the clockwise Plant X orbit, from its point of perihelion nearest our Sun, out to aphelion, its furthest distance from the Sun and then back again.

In this case, this glyph shows us exactly where Planet X will be relative to our Sun in December 2012.

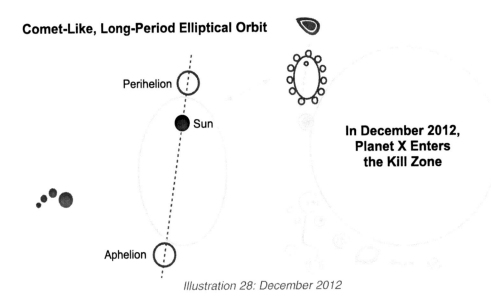

Illustration 28: December 2012

This glyph shows an alignment of the Sun and Planet X where a straight line can be drawn through the Sun in the upper interior of the glyph and the object's points of perihelion and aphelion.

In December 2012, Planet X will begin its transit from its point of perihelion to the ecliptic (the plane of the solar system) which will be the most violent period of the flyby. However, the worst of this period will pass more quickly than simple math would suggest.

This is because the closer Planet X comes to the Sun, the steeper its angle of attack becomes; therefore, its velocity will increase. A good way to imagine this is with a popular amusement park ride called The Whip.

Like objects in long-period elliptical orbits, the gaily painted cars of The Whip trundle between two great turns along boring straightaways connecting the two turns. As the car approaches a turn, it will suddenly speed up and then whip around in the opposite direction. The same can be imagined with Planet X whipping about its own points of perihelion and aphelion.

At least in some small measure, celestial mechanics could help mitigate the duration of this tribulation, but then again, there's still the worst of it.

The Kill Zone

Future generations will look back on December 21, 2012, much as we look back on dates like September 11, 2001, a date that launched us into the longest war in our nation's history.

They will create their own names for this dark period of global tribulation for it will be felt the world over and expressed in many tongues. However, for those who are about to become the grist of this evolutionary wheel, it will simply be known as the kill zone, and our moon will point the way.

A Waning Crescent Moon

With Avebury 2008, the crop circle makers used a lunar phase as the time stamp for when we will all see Planet X as it enters the kill zone. This is when we will begin to see the worst of the solar violence.

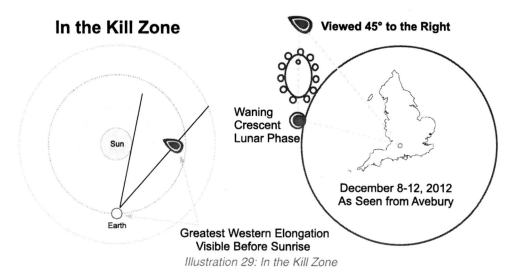

Illustration 29: In the Kill Zone

The lunar glyph appended to the large empty ring shows the moon in a waning crescent phase (from December 8-12, 2012). During this time, the moon will move through Libra, Scorpio, and Sagittarius.

This is important to note, as the celestial alignment for December 21, 2012, puts the Sun in Ophiuchus.

Consequently, it is totally consistent with the lunar cycle shown in the second part of the Avebury formation, as Ophiuchus is found between and just above Sagittarius and Libra.

At this point, the only relevant question is: When do we go to ground? If you're waiting to ask that question, that is your choice. However, if you wait, those who could competently answer the question will have already disappeared.

Shelter by Day – Forage by Night

As Planet X transits the kill zone, every living soul on the planet will see what appears to be two suns in the sky—one sun and its smaller, but blazing twin. This will be a time when the unprepared will hide underground as best they can during the day.

In the relative safety of night, they will forage for water, food, and supplies. These will be scarce, so it will be a time of great danger for the unprepared as Planet X transits the kill zone.

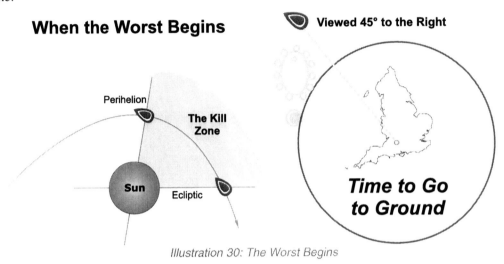

Illustration 30: The Worst Begins

In more specific terms, the kill zone is the end of the beginning. It occurs when Planet X passes its point of perihelion as shown in the Avebury 2008 formation and speeds toward the ecliptic. Conversely, the beginning of the end is when Planet X reaches the ecliptic—that point in its orbit where it crosses the plane of the solar system. After this, the perturbations will start to abate.

However, while Planet X is in the kill zone, interactions between it and the Sun will be extremely violent. We can expect to see severe storms and cosmic sprites, or what is known as solar lightning.

Sprites can reach out through space and, with a cosmic snap, tear open great canyons on a planet's surface—canyons as great as the Grand Canyon on Earth and the Valles Marineris on Mars.

Given that this energetic duel in the sky will be so devastating, we must ask: Is this survivable?

The answer is yes, but it will not be easy. In this regard, the message of the crop circle makers is very clear. If you're waiting to see what happens on December 21, 2012, it will already be too late for to prepare. *Start preparing now!*

For the unprepared, the trigger event will be the worst outcome of the kill zone transit. It will be the solar event that raises the temperature of Earth's core, thereby triggering a pole shift event months later. Like a slow-motion train wreck, the culminating event will be the type of pole shift predicted by Edgar Cayce.

Trigger Event

One of the most remarkable aspects of the Avebury formation is the precise manner in which the second part was partially overlaid on the first. While this is an example of an impressive capability, it nonetheless carries the most ominous prediction of all.

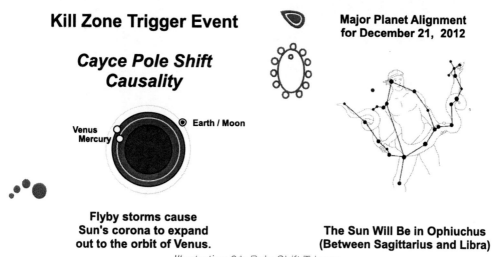

Illustration 31: Pole Shift Trigger

In the first part, we see a normal representation of the Sun. However, with the addition of the second part, the Sun's corona pushes outward, engulfing all of Mercury and extending to the orbit of Venus.

In other words, the radiation and gases generated by solar interactions with Planet X will slam deep into the Earth, heating the planet's core much like a tea kettle on a stove. However, when we hear the whistle of this tea kettle, it will not be teatime; it will be pole shift time.

That is when humanity will begin crossing the cusp. This event will claim more lives than any other event throughout the entire tribulation. And before that, we will also see a replay of the Ten Plagues of Exodus, and ours will be a similar suffering.

4

The Dragon's Tail

The crop circle makers have accurately depicted how Planet X appears when it is entering the kill zone. At that time, every soul on the planet will see it. However, what will become progressively more difficult over time will be our ability to see the dust tail that trails behind Planet X in the object's own trajectory.

While the Avebury 2008 formation shows us a narrow span of time, if we project that event two years into the future we see the very real possibility that after the solar storms and sprites of the kill zone, we'll also get slammed by Planet X's tail as well, and not just once, but twice!

Imagine a mile-long freight train passing you at night during a full moon. The locomotive engines pulling the train are so brightly lit they cannot help but catch your eye. Directly behind them are brightly painted cars that easily stand out in the moonlight.

However, behind the brightly painted cars and the brightly lit locomotive engines is more than a half mile of darkly painted freight cars. Starting with a deep shade of gray, they progressively darken until the cars at the back of the train are pitch black.

The difference between these rail cars is not what happens when you drive across the railroad tracks thinking all is clear. The cars painted pitch black are just as big and just as devastating as any other in the train. Rather, the difference is what you can see and when, and this matter was determined nearly a decade ago by a knee-jerk political reaction.

What You Can't See Won't Hurt You

A constant thorn in humanity's collective consciousness is our inbred fear of comets. Comets were a rich feeding ground for yellow journalism, especially in the 1990s after the Internet made first-time Near Earth Orbit (NEO) sightings more accessible. Newspapers (and most particularly English tabloids) monitored these same data so as to profit from the fear generated with fearmonger headlines.

When astronomers reported observations of a new object in an Earth-crossing orbit with the slightest chance of a future impact event, headlines quickly roared out end-of-the-world predictions. It was easy money for the newspapers and tabloids because, as further observations were made, the odds of an impact inevitably dwindled—a postscript that arrived long after the profits were realized.

However, those initial headlines did cause needless panic and, tired of dealing with the public concern generated by these premature impact event headlines, NASA's solution was to throw free speech under the bus. In response to its annoyance with this new Internet-driven access to information, NASA implemented a new embargo policy on all new NEO observations.

In an Associated Press article dated April 15, 2001, this was called the "doom's day news protocol." As a political invention, the protocol employs a "what you can't see won't hurt you" approach. Since then, no new observations have been made public without NASA's blessing, and the consequences for those who violate this rule and their institutions are severe.

Hence, this NEO embargo policy has been incredibly effective at throttling what had been a free and unfettered exchange of discovery. The result is subtle but nonetheless measurable.

Irresponsible 1990-style headlines about asteroids hitting the Earth have been replaced with a steady stream of after-the-fact headlines telling us about newly discovered NEO objects that just happened to have passed close to the Earth, a few days prior to the announcement.

The result is that while NASA's heavy-handed approach is of clear benefit to the agency's own public relations efforts, it nonetheless has throttled the reporting of NEO discoveries. Obviously, it is much easier to throttle astronomers than reckless journalists.

The result is NASA's power in this regard is both entrenched and scalable. Having enjoyed a decade of success with its ability to throttle the reporting of NEO discoveries, the popular assumption that this agency would cast its success aside for an object like Planet X is Pollyannaish at best. The nagging hope that the government would just have to be honest about something this big is so naïve that even Hollywood knows better.

In 1998, Hollywood released two hugely popular asteroid-impact disaster movies, *Deep Impact* and *Armageddon*. In both films, the government withholds any announcements until the last possible moment, ostensibly so it can effect its own solution without being hampered by public concerns.

Slaves to what sells, Hollywood presented the government this way in both films because this is what moviegoers would expect. It was a believable scenario, even if it was only for entertainment. Yet, when many people step outside the theater, they assume that the box office does not drive actual reality.

Step forward to the next major space-disaster movie—Sony's *2012* (2010) starring John Cusack. In the film, director Roland Emmerich gives us very realistic insight into how things actually work today and into the future.

To see this subtext, you must view the film a second or third time so that you're no longer distracted by the eye candy of special effects. Do that and you will see a brutal and honest depiction of things as they truly are.

In the movie, elites and government leaders are focused solely on securing their own personal safety and maintaining dominance on the planet. To do this, they actively suppress knowledge of the impending disaster until they've made good their escape. Consequently, the rest of humanity comes to learn of the catastrophe only as they die from its effects.

This subtext in the film is not about sales at the box office. It is Roland Emmerich's prescient warning to us all: Unless we are chosen or can afford passage, we will be betrayed and thrown away.

Consequently, when we all see what appears to be a bright comet in the sky, expect false assurances designed to prey upon the naïve hopes of humanity.

When that comet lights up so brightly that it appears like a smaller second Sun, it will be too late for those who've invested in naïve hopes. As this object depicted in the second part of the Avebury 2008 formation forces us to face the truth, those who have withheld it from us for years will already be elsewhere, on this world or another.

Meanwhile, it is vital that we remember that this object has a massive dust tail trailing behind it in the same orbit.

The Dust Tail

The Avebury 2008 formation depicts a comet-like object rising upward through the core of our solar system from deep within our southern skies. This steep southern approach of Planet X brings it through an area of the sky that is difficult to observe until it is inside the orbit of Jupiter, where it begins to interact with our Sun. At this time, we will see the dust tail.

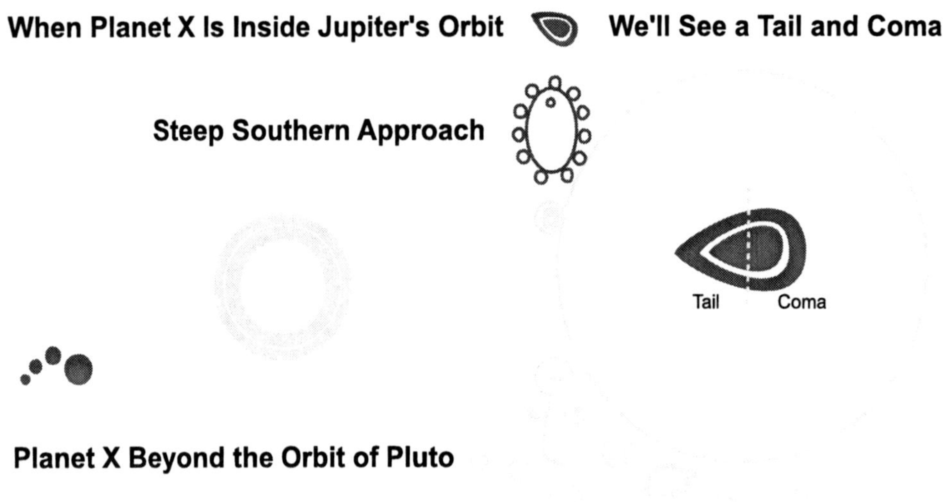

When Planet X Is Inside Jupiter's Orbit   **We'll See a Tail and Coma**

Steep Southern Approach

Planet X Beyond the Orbit of Pluto

Illustration 32: Tail and Coma

When the orbit of a comet brings it inside the orbit of Jupiter, it is warmed by the Sun enough to form a horseshoe-shaped coma on the leading edge. It also forms two general types of tails. The most recognizable type is ionized gas, and it always points away from the sun.

The second type, the dust tail, is of particular concern here. Dust tails trail behind the object in its own orbit around our Sun. In the case of a massive object such as Planet X, this dust tail will be orders of magnitude wider and longer that those of comets such as Hale-Bopp. This is because this object will be big enough to appear as a second Sun in the sky.

It Will Appear as a Smaller Sun

Once Planet X is in the kill zone, the same forces that caused it to form a coma and ionized gas tail will increase tremendously. This interaction will cause Planet X to appear as a second Sun in the sky. What will we see and how will it affect us? A realistic glimpse of the future was painted by Veronica Lueken, a Roman Catholic mystic.

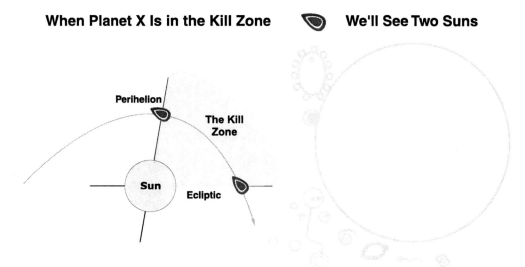

Illustration 33: We See Two Suns

Claiming to have experienced apparitions of the Virgin Mary, as well numerous other Catholic saints, Luken made several predictions about the day we would all see two Suns in the sky. Of her many prophecies, the most relevant (if not chilling) was given in 1976:

```
"Oh, my! Our Lady is pointing up to the sky, and I
can see a tremendously large BALL OF FIRE. It's
beyond description. It's the most frightening
thing I've ever seen. And it's going through the
air at a very fast speed, and I can see over on
the right side of the sky an outline of the earth.
And the ball is heading for the earth and it's
striking, the tail is setting fire to the side, I
can see here, of the earth. Oh! Oh! The tail has
intersected the earth, and the ball is now
circling the earth. Oh! Now it's growing very
dark. I can feel the great heat. Ohh!" Veronica
Lueken: Aug. 21, 1976
```

What she is confirming in the Avebury 2008 formation is that when Planet X is in the kill zone, the interaction between it and our Sun will be so severe that Planet X will become bright enough to appear as a second, smaller Sun in the sky—a "BALL OF FIRE."

It should also be noted that in 1986, Lueken's prophecies were "officially" dismissed as being "contrary to the teachings of the Catholic Church." Still today though, 15 years after her passing in 1995, the predictions of this Catholic seer from Bayside, New York, still command a loyal following. Her followers are clearly immune to the dissuasions of the Church.

Let us keep that in mind when those in positions of trust and power try to assuage our fears with dismissive explanations. Perhaps the most memorable line from the *2012* movie: "When *They Tell You* Not To Panic... That's When You Run!"

This is because this flyby will begin with a horrible period of solar violence, followed by waves of meteorite violence as our planet transits the long dust trailing behind Planet X. Not once, but possibly twice. Therefore, we must wonder how big such a destructive force can be. The initial point of truth for that answer is the nature of the object itself. More to the point, has anyone seen it? Yes, they have.

The Nature of the Beast

One of the things that newcomers to the topic often do is try to find Planet X using one of the popular Internet sky programs. But newcomers do not understand that these visual databases are just as subject to disinformation as any other information source. Newcomers nonetheless assume that something of this magnitude would automatically demand immediate and full disclosure.

However, one can find a few crumbs lying either on the floor or on the table and one such crumb is called the South Pole Telescope (SPT) at the South Pole.

The South Pole Telescope

The South Pole Telescope is a massive infrared telescope, and its stated mission could be achieved with equal results in a far more accessible location and at a significantly lesser cost. Yet it was placed at one of the most hostile and inaccessible points on the globe. Why?

In terms of visible light observations, Planet X is still well beyond the orbit of Jupiter deep in our southern skies. This makes it a dust-shrouded needle in a stellar haystack.

However, when observed along the infrared spectrum, Planet X has a rather large signature. This makes the SPT the best telescope for observing a dusty object with a strong infrared signature as it rises upward through our southern skies toward the core of our solar system.

Yowusa.com first broke the SPT story in 2006. Then in the first half of 2008, three SPT disclosure videos appeared on Youtube.com with images of Planet X, alleged to be actual SPT images of the object. Although the Internet is rife with hoax images (many which are planted to created confusion), these three contain the only images ever authenticated by this author.

This is because of two factors. First, the images in all three videos held up to close scrutiny, but the second factor was even more compelling. It was the rapid onset and mean-spirited disinformation response that ensued, to include removal of the videos by Youtube.com and account closures.

With Planet X evidence, always remember that the closer it comes to the truth, the faster it disappears.

NibiruShock2012 and DNIr4808n Disclosure Videos

The first two videos were published by a Youtuber with the screen name, NibiruShock-2012. He posted his first two videos in January and February 2008. After that, his account was stolen by disinformationists and trashed with hoax notices.

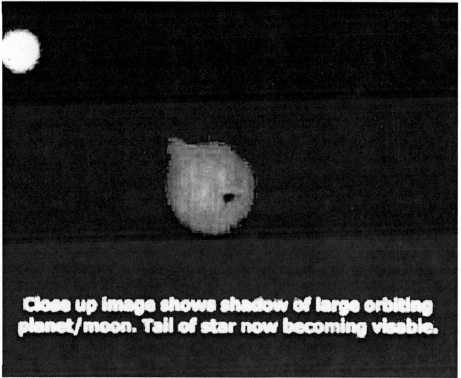

Illustration 34: NibiruShock2012 - Jan/Feb 2008

It is true that you can retrieve a stolen Youtube.com account—that is, provided you first identify yourself. Obviously, this is not an acceptable option for an anonymous whistleblower. For the record, the real NibiruShock2012 did contact this author prior to the theft of his YouTube account. Following the theft, I have not heard from him.

The third video was posted in September 2008 by a second Youtuber, with the screen name DNIr4808n.

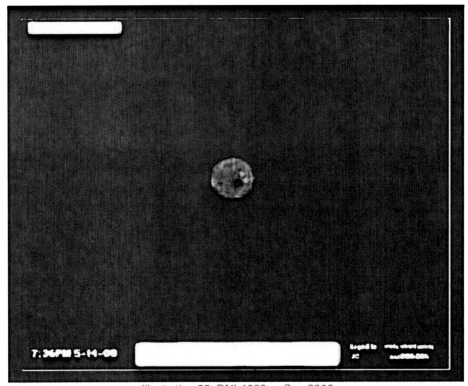
Illustration 35: DNIr4808n - Sep 2008

The video was up for little over a day because DNIr4808n named names. Consequently, Youtube.com summarily removed his video and terminated his account with uncommon speed.

Thanks to a hot tip from a reliable source, this author was able to download a copy of the DNIr4808n video before Youtube.com removed it and terminated the account.

What can be said about the DNIr4808n video in general is that the text was very sincere. It also exhibited an excellent understanding of key Planet X orbital concepts.

What the Disclosure Videos Revealed

The images by both NibiruShock2012 and DNIr4808n show a reddish dusty object with a substantial dust tail along with several satellites (planets, moons, etc.). In other words, they show a mini-constellation with a brown dwarf sun as a smaller twin to our own Sun.

NibiruShock2012

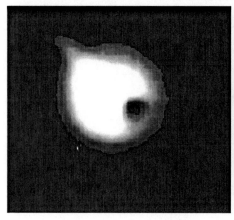

JAN/FEB 2008

DNIr4808n

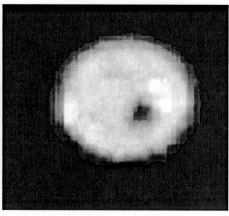

SEP 2008

Illustration 36: SPT Leaks Composite

A brown dwarf is also called a "failed star" because it has enough mass to ignite. This ignition creates a large dusty disk which will persist; however, the star's mass is insufficient to sustain nuclear fusion in its core.

Nonetheless, a failed star will continue to give off ample heat, much like a barbecue charcoal briquette does, after first being lit. Therefore, what is of real concern to us is all of the dust that surrounds it and how wide a path this dusty disk cuts through space.

Unlike a comet, such as Hale-Bopp, the dust tail of an object the size of a mini-constellation will be massive; therefore, it does not necessarily need to have a precise Earth-crossing orbit. As the old saying goes, "close enough for government work."

To illustrate the point, assume that two people are shooting guns at you. The first is called Hale-Bopp and he is shooting a .22 rife, such as we find in the shooting gallery at a circus arcade. The other calls himself B-52, and what he sends our way will be like a flying maelstrom of exploding cluster bombs. It will be remembered as it was predicted for centuries as "the dragon's tail."

The Dragon's Tail

What the NibiruShock2012 and DNIr4808n disclosure videos showed us is that Planet X is a mini-constellation with a brown dwarf companion to our own Sun. This description fits with the dragon prophecy of British seer Ursula Southeil (1488–1561). Known as Mother Shipton, her prophecies first appeared in print in 1641, 80 years after her death.

In those prophecies, she predicts that Earth will fly through a Planet X maelstrom. The most accurate copy of her prophecies was published by "Nexus Magazine," (March 1995, Volume 2, # 24). Shipton's predictions for these times are as follows:

Illustration 37: Mother Shipton

A fiery dragon will cross the sky
Six times before this earth shall die.
Mankind will tremble and frightened be
for the sixth heralds in this prophecy.

For seven days and seven nights
Man will watch this awesome sight.
The tides will rise beyond their ken
To bite away the shores and then
The mountains will begin to roar
And earthquakes split the plain to shore.

And flooding waters, rushing in
Will flood the lands with such a din
That mankind cowers in muddy fen
And snarls about his fellow men.

He bares his teeth and fights and kills
And secrets food in secret hills
And ugly in his fear, he lies
To kill marauders, thieves and spies.

> Man flees in terror from the floods
> And kills, and rapes and lies in blood
> And spilling blood by mankind's hands
> Will stain and bitter many lands
>
> And when the dragon's tail is gone,
> Man forgets, and smiles, and carries on
> To apply himself - too late, too late
> For mankind has earned deserved fate.
>
> His masked smile - his false grandeur,
> Will serve the Gods their anger stir.
> And they will send the Dragon back
> To light the sky - his tail will crack
> Upon the earth and rend the earth
> And man shall flee, King, Lord, and serf.
>
> But slowly they are routed out
> To seek diminishing water spout
> And men will die of thirst before
> The oceans rise to mount the shore.
>
> And lands will crack and rend anew
> You think it strange. It will come true.

Two specific predictions directly mention the "dragon's tail" and at two different times. The first mention places this prediction in the period just following the transit of Planet X through the kill zone. At that point, the object is leaving the core of the system.

> And when the dragon's tail is gone,
> Man forgets, and smiles, and carries on

From this point, an unspecified interval of time has passed. However, it need only be a matter of months before the uninformed and unprepared mistakenly assume that the danger has passed.

> And they will send the Dragon back
> To light the sky - his tail will crack
> Upon the earth and rend the earth

At this point, we can build a dust tail scenario using the Avebury 2008 formation, Veronica Lueken's two Suns in the sky prediction, and Mother Shipton's two separate dragon's tail prediction.

Starting From Avebury

This scenario begins with the Avebury 2008 formation. It tells us where to stand, when, in which direction we need to look, and for what. Therefore, the scenario begins in early December 2012 while the moon is seen in a waning crescent phase from the Avebury site.

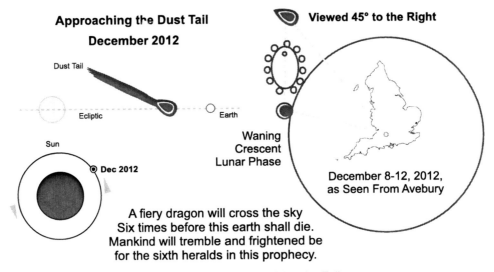

A fiery dragon will cross the sky
Six times before this earth shall die.
Mankind will tremble and frightened be
for the sixth heralds in this prophecy.

Illustration 38: Approaching the Tail

Planet X will be seen 45° to the right from Avebury, and we will also see its dust and gas tails. The dust tail is the critical issue in this scenario because the formation shows us it will be a large object. Therefore, it will have a correspondingly large and long dust tail behind it.

Consider this. In April 2010, Iceland's Eyjafjallajökull volcano erupted for six days. This was a first-of-a-kind event, though relatively small when compared with other eruptions. However, the impact on air travel across western and northern Europe was staggering. Well over 100,000 flights were canceled at a cost to the airline industry of some $200 million, according to the International Air Transport Association (IATA).

When we transit the dust tail behind Planet X, every airplane in the world will be grounded and much worse. Another thing to keep in mind is that Planet X, as seen from Avebury, is in a clockwise orbit around the Sun, whereas the Earth is in a counterclockwise orbit around the Sun.

This means that one is not chasing or leading the other, which would be preferable to what Avebury 2008 shows us—that Earth and Planet X are traveling toward each other like two aircraft blindly converging toward a mid-air collision point.

Given this convergence plus that fact that Planet X will still be carrying extra speed from its perihelion whip about the Sun, events will move quickly toward the next catastrophe in the first quarter of 2013.

Earth's First Transit of the Dust Tail

A comet's dust tail follows the same orbital path as the comet, and a failed star like Planet X will have a very long tail filled with the dust and debris of its own ignition, plus whatever it picks up along the way. As we fly through the dust tail maelstrom in 2013, what could we experience?

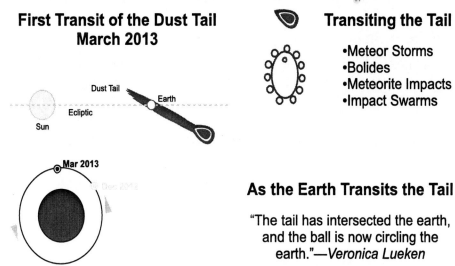

**First Transit of the Dust Tail
March 2013**

Transiting the Tail

- Meteor Storms
- Bolides
- Meteorite Impacts
- Impact Swarms

As the Earth Transits the Tail

"The tail has intersected the earth, and the ball is now circling the earth." —*Veronica Lueken*

Illustration 39: First Transit 2013

For certain, we'll see meteor storms and bolides tearing across our skies and with horrifying sounds. In *The Kolbrin Bible,* we're told that this frightened the ancients so much that men became impotent and women infertile. Or, in other words, the horror was so great that it shut down one of our most primal drives—reproduction.

However, more related to this scenario is a description from Exodus. In particular, the meteorite impacts and showers will spawn a series of interlinked catastrophes, as described in the Seventh Plague of Exodus:

> **Exodus 9:23-24**
>
> Then Moses stretched out his staff toward heaven, and the LORD sent thunder and hail, and fire came down on the earth. And the LORD rained hail on the land of Egypt; there was hail with fire flashing continually in the midst of it, such heavy hail as had never fallen in all the land of Egypt since it became a nation.

The key point here is that we, as did the ancient Egyptians, will experience completely new forms of meteorite impacts and showers. This alone helps explain fear-driven impotency and infertility.

It is bad enough to see something familiar and horrible and quite another to see something terrible for the very first time. In times like these, people will cling to the thinnest threads of hope and this will make them susceptible to the false grandeur of others.

The Price of False Grandeur

As Planet X travels to and from the furthest regions of our solar system, it scoops up everything in its path, much like a cosmic riding mower with a rear clipping bag that goes on and on. What is in the bag will be our problem, so to speak.

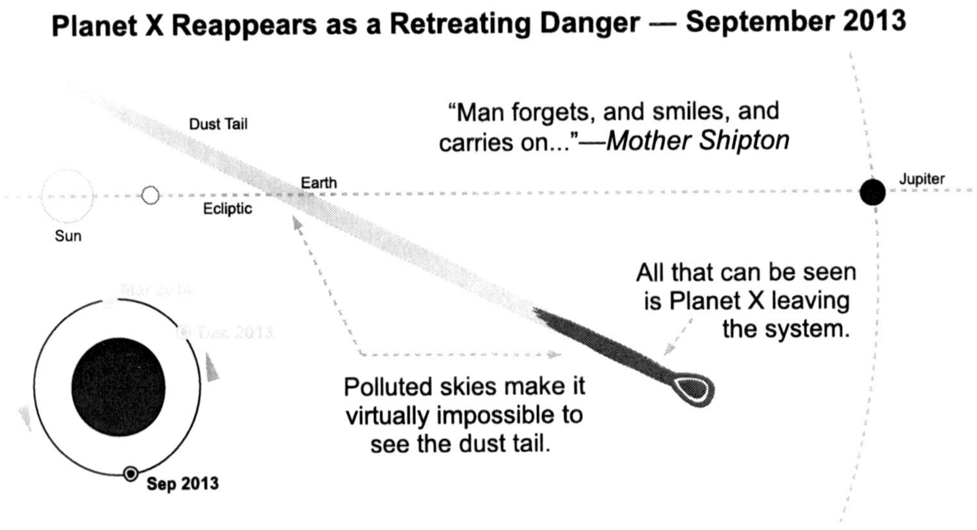
Planet X Reappears as a Retreating Danger — September 2013
Dust Tail
"Man forgets, and smiles, and carries on..."—*Mother Shipton*
Earth
Jupiter
Ecliptic
Sun
Mar 2014
Dec 2013
Sep 2013
All that can be seen is Planet X leaving the system.
Polluted skies make it virtually impossible to see the dust tail.

Illustration 40: Planet X Reappears

Gone will be our beautiful blue skies. After hundreds of meteorite impacts, volcanic eruptions, and raging fire storms, they will black-streaked and reddish-orange for months on end.

In fact, over the last decade, this author has received personal accounts from unrelated people from every walk of life and from all four quarters of the globe.

There are several themes to these reports, such as two Suns in the sky and tsunamis. However, black-streaked and reddish-orange sky visions and dreams account for well over half the reports. This is a consciousness event, which is not a time for dismissive, educated guesses. Rather, we would do well to take Mother Shipton's warning to heart:

```
His masked smile - his false grandeur,
Will serve the Gods their anger stir.
And they will send the Dragon back
```

The warning is clear. Do not allow complacency and arrogance to woo you into a state of "false grandeur," for it will be your undoing.

Second Transit of Planet X Dust Tail — March 2014

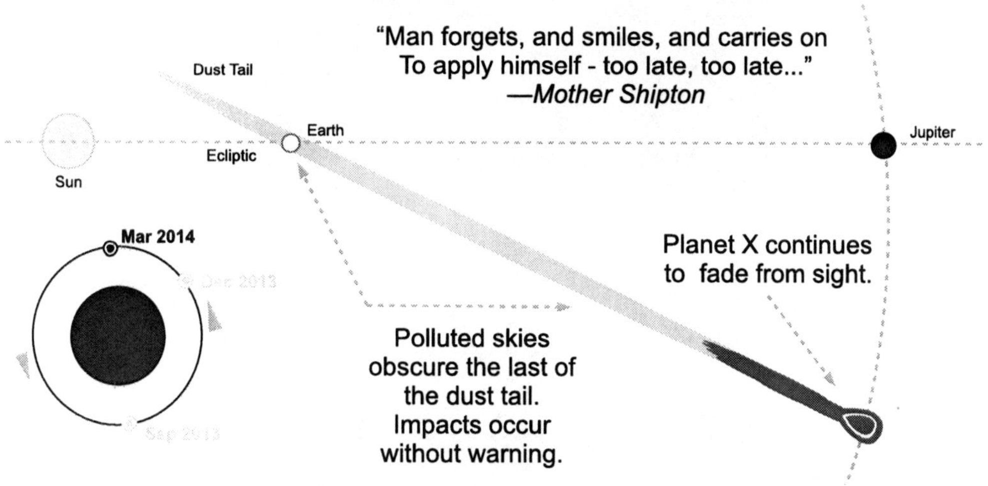
"Man forgets, and smiles, and carries on
To apply himself - too late, too late..."
—Mother Shipton
Dust Tail
Earth
Jupiter
Ecliptic
Sun
Mar 2014
Planet X continues to fade from sight.
Polluted skies obscure the last of the dust tail. Impacts occur without warning.

Illustration 41: Second Transit 2014

Rather, if you survive the first transit of the tail, immediately begin preparations for the second transit as it will be as bad or perhaps even worse. A flyby such as this is documented by the ancient Egyptians.

The Last Flyby as Told by the Ancient Egyptians

Mother Shipton began her dragon prediction by stating that Planet X has flown through the core of our solar system before.

```
A fiery dragon will cross the sky
Six times before this earth shall die.
Mankind will tremble and frightened be
for the sixth heralds in this prophecy.
```

In his book *Planet X and The Kolbrin Bible Connection*, author Greg Jenner offers proof from *The Kolbrin Bible* and several other ancient wisdom texts and folklore to show that previous Planet X flybys caused the sinking of Atlantis, the Great Deluge (Noah's flood), and the Ten Plagues of Exodus.

Illustration 42: Planet X - Kolbrin Bible

Mother Shipton's warning says the "Dragon" will cross the sky a sixth time. While it is difficult to understand how she arrived at this number, the fact remains that the number is greater than one. Ergo, the point that this object has been here before is sufficient to the task at hand. Reconstructing the sequence of catastrophes this object created in the Exodus account:

> **Exodus 9:24**
>
> ...there was hail with fire flashing continually in the midst of it, such heavy hail as had never fallen in all the land of Egypt since it became a nation.

The Exodus account describes the event from the Hebrew perspective and uses allegory to document the history of the Hebrews and their right as a nation to exist. It is why each year Jews gather in the spring to remember the Exodus story with the Passover Seder.

The *The Kolbrin Bible* presents the Egyptian perspective, which features significantly greater detail. As the Egyptians were the losers, their account was written in a straightforward way, very much like a man-on-the-street television interview. What they tell us is that they were bewildered that the god of the Hebrews bested their entire pantheon of gods.

The Egyptians' account says that the Pharaoh of Exodus personally led the attack against the host after Moses had crossed to the other side and slaughtered a great many of them, but then drowned when the waters returned.

The successor to the Pharaoh of Exodus later ordered work to begin on what was named the Great Book. It is a secular work and serves essentially the same function as the the 9-11 Commission report. The attempt was to understand how a small group of slaves could best an empire.

What remains of those accounts is now contained in the Egyptian texts of *The Bronzebook*, the first six books of *The Kolbrin Bible*. The Egyptian account of Exodus is found in the fifth book, the Book of Manuscripts. It gives us a very prescient clue as to what actually caused the original Ten Plagues of Exodus, with details that lead to a more scientific understanding of the event.

Meteorites, Algae Blooms, and Microcystin

As you read this account, keep the following in mind. Planet X will be dragging a tremendous number of meteorites with it, which are basically classified as iron or stony and with various sub-groups. However, keep in mind that both the Egyptian and Hebrew accounts of Exodus tell us that these meteorites were of a type never seen before.

Most likely, there were plenty of the iron and stony meteorites we're accustomed to seeing these days, but there would also have been these anomalous meteorites. Together, they would have formed a real toxic brew of death from the skies.

The iron meteorites would have caused the waters to turn blood red, for the same reason brickyards add iron to the mix when making red house bricks. The iron gives them that deep red color. However, these iron meteorites would have also contained s chreibersite, a rare iron nickel phosphide mineral.

During a meteorite storm, vast amounts of schreibersite would impact the soil and waters of the Earth, and this is critical to understanding the Egyptian account of Exodus.

This is because the phosphorus contained in meteorites bearing schreibersite is a necessary and vital element for life on this planet. In the case of algae, the phosphorus contained in the chreibersite becomes a banquet for blue-green algae and creates blooms.

With their growth unchecked, these algae blooms can then produce a toxin called microcystin, which can persist in the water long after the blooms disappear. A hepatotoxin microcystin can affect the liver and will make you sick from direct contact, swallowing, or via airborne droplets.

The principal symptoms of microcystin poisoning can begin to appear within a few hours and include:

- Skin contact. Rash, hives, and or skin blisters. The lips are especially sensitive.
- Swallowing. Pain, nausea, vomiting, diarrhea, severe headaches, and fever.
- Inhaling. Runny eyes and nose, cough, and sore throat, chest pain, asthma-like symptoms, or allergic reactions.
- High exposure. High levels can cause permanent liver damage.
- Fatal exposure. In sufficient amounts microcystin can cause death for all manner of life, including humans and their domesticated animals.

With this in mind, we turn to the following excerpt from the Egyptian account of Exodus, from the Book of Manuscripts:

```
The Kolbrin Bible
Book of Manuscripts 6:11-24
```

```
MAN:6:11 Dust and smoke clouds darkened the sky
and coloured the waters upon which they fell with
a bloody hue. Plague was throughout the land, the
river was bloody, and blood was everywhere. The
water was vile and men's stomachs shrank from
drinking. Those who did drink from the river
vomited it up, for it was polluted.
```

```
MAN:6:12 The dust tore wounds in the skin of man
and beast. In the glow of the Destroyer, the Earth
was filled with redness. Vermin bred and filled
the air and face of the Earth with loathsomeness.
```

```
MAN:6:13 Trees throughout the land were destroyed
and no herb or fruit was to be found. The face of
the land was battered and devastated by a hail of
stones, which smashed down all that stood in the
path of the torrent. They swept down in hot
showers, and strange flowing fire ran along the
ground in their wake.
```

MAN:6:14 The fish of the river died in the polluted waters; worms, insects and reptiles sprang up from the Earth in huge numbers. Great gusts of wind brought swarms of locusts which covered the sky. As the Destroyer flung itself through the Heavens, it blew great gusts of cinders across the face of the land. The gloom of a long night spread a dark mantle of blackness, which extinguished every ray of light. None knew when it was day and when it was night, for the sun cast no shadow.

MAN:6:15 The darkness was not the clean blackness of night, but a thick darkness in which the breath of men was stopped in their throats. Men gasped in a hot cloud of vapour, which enveloped all the land and snuffed out all lamps and fires. Men were benumbed and lay moaning in their beds. None spoke to another or took food, for they were overwhelmed with despair. Ships were sucked away from their moorings and destroyed in great whirlpools. It was a time of undoing.

MAN:6:16 The Earth turned over, as clay spun upon a potter's wheel. The whole land was filled with uproar from the thunder of the Destroyer overhead and the cry of the people. There was the sound of moaning and lamentation on every side. The Earth spewed up its dead, corpses were cast up out of their resting places and the embalmed were revealed to the sight of all men. Pregnant women miscarried and the seed of men was stopped.

MAN:6:19 On the great night of the Destroyer's wrath, when its terror was at its height, there was a hail of rocks, and the Earth heaved as pain rent her bowels. Gates, columns and walls were consumed by fire, and the statues of gods were overthrown and broken. People fled outside their dwellings in fear and were slain by the hail. Those who took shelter from the hail were swallowed when the Earth split open.

MAN:6:21 The land writhed under the wrath of the Destroyer and groaned with the agony of Egypt. It shook itself and the temples and palaces of the nobles were thrown down from their foundations. The highborn ones perished in the midst of the ruins, and all the strength of the land was stricken. Even the great one, the first born of Pharaoh, died with the highborn in the midst of the terror and falling stones. The children of princes were cast out into the streets and those who were not cast out died within their abodes.

MAN:6:22 There were nine days of darkness and upheaval, while a tempest raged such as never had been known before. When it passed away, brother buried brother throughout the land. Men rose up against those in authority and fled from the cities to dwell in tents in the outlands.

MAN:6:24 The slaves spared by the Destroyer left the accursed land forthwith. Their multitude moved in the gloom of a half dawn, under a mantle of fine swirling grey ash, leaving the burnt fields and shattered cities behind them. Many Egyptians attached themselves to the host, for one who was great led them forth, a priest prince of the inner courtyard.

It is interesting to note that Moses never mentions the name of the Pharaoh of Egypt in the Hebrew accounts and neither do the ancient Egyptians in the Book of Manuscripts. In fact, they only refer to him as "one who was great led them forth, a priest prince of the inner courtyard."

As a priest prince of the inner courtyard, Moses would have been privy to state secrets, including accounts of previous flybys of Planet X, and would have known the chaos of such an event would be the opportunity for him to lead the Hebrews out of bondage.

He would also have known that unlike the flyby that caused the Great Deluge (Noah's flood), the flyby of Exodus would not be as severe.

However, for those of us in the present, the crop circle makers of the Avebury 2008 formation have sent a different message: This flyby can cause a pole shift that, unlike the other calamities described in this chapter, future generations will remember as the single greatest die-off event of the coming flyby. The Egyptians called it The Great Winnowing.

5

The Great Winnowing

The Great Winnowing, as it is called by the ancient Egyptians, will be remembered as the last and greatest tribulation, and it will begin with a pole shift—the very one predicted by the most documented of the 20th century, Edgar Cayce, "the sleeping prophet."

> **Edgar Cayce: Reading No. 3976-15**
> **New York City, NY - January 19, 1934**
>
> ```
> "...There will be upheavals in the Arctic and in
> the Antarctic that will make for the eruption of
> volcanoes in the Torrid areas, and there will be a
> shifting of the poles... when HIS LIGHT [the
> Christ Star / Wormwood] WILL BE SEEN AGAIN in the
> clouds."
> ```

During the period Planet X transits the kill zone, the effects of solar eruptions and meteorite impacts will resonate with a catastrophic synergy, lasting long enough to unleash the forces for a Cayce pole shift.

From first rumble to last rumble, the event will take approximately a year, though the most intense part will take a matter of days.

At first glance, it is difficult to imagine a single event so powerful that it initiates a year-long process (or more) wherein the lithosphere (the crust and upper mantle of the earth) eventually twists the Earth's core into a new orientation. Frankly, though, it is a logical response that works.

This is because the chance of a single cause is remote whereas an unbroken string of catastrophic events culminating in a pole shift event is the more likely explanation. Furthermore, it is consistent with the message of the Avebury 2008 Planet X formation.

Although the crop circle formation focuses on the month of December 2012, by trending those predictions forward in time, one can easily see a chain of catastrophic events anchored at one end by the arrival of Planet X in the core of our solar system and at the other by a Cayce pole shift.

However, before we begin building links in this chain of catastrophe, we must ask the most elemental question: Is a pole shift actually possible?

Is a Pole Shift Actually Possible?

Despite Velikovsky's intellectual mauling by the scientific community, he did succeed in one regard. He successfully injected the possibility of a pole shift into the public dialog.

The man who put more acceptable flesh on the battered bones of Velikovsky's premise was Charles Hapgood in his book, *The Earth's Shifting Crust* (1958).

A respected scientist, Hapgood's shift theory was inoculated from the abuse suffered by Velikovsky by none other than Albert Einstein.

Einstein and Velikovsky corresponded frequently, which makes it obvious that the two shared mutual interests. After witnessing the medieval abuse meted out to a friend, Einstein did not back away. Rather, he wrote the forward to *The Earth's Shifting Crust*, and in it addressed the issue of pole shift feasibility straight on:

```
Earth's Shifting Crust: A Key to Some Basic
Problems of Earth Science
Charles H. Hapgood
Pantheon Books Inc., 1958


FOREWORD by Albert Einstein
```

> Without a doubt the earth's crust is strong enough not to give way proportionately as the ice is deposited. The only doubtful assumption is that the earth's crust can be moved easily enough over the inner layers.

The possibility of a pole shift event is, as Einstein carefully phrased it, a "doubtful [an uncertain outcome] assumption [hypothesis]." The message to his peers was simple. This hypothesis could eventually prove to be correct. Likewise, it could fall on its face.

Therefore, both possibilities must be given weight (i.e., pole shifts can happen), but could Einstein have been wrong? Even Einstein answers that as well. Who is to say.

Nonetheless, he was a visionary with a genuine love for humanity and so warned us that: "If the bee disappears from the surface of the earth, man would have no more than four years to live. No more bees, no more pollination ... no more men!" As fate would have it, the bees are disappearing. Let us not be so arrogant that we join them.

The pole shift will be a very personal matter for each of us, wherein the choices we make as individuals will decide our outcomes in what shall be a time remembered as crossing the cusp. A brief, geological moment of massive evolutionary change that reorders life on this planet.

In this time our modern but unsustainable world will end and those who endure will enter into the next and greatest epoch of humanity's journey through time, if they choose it to be. So then, who are these few?

The Search for a Cusp Survival Paradigm

We live in a world of logical materialism, a more pleasant-sounding oxymoron for plain old, dog-eat-dog greed and exploitation—a world that rewards service to self, though the relative morality of that is a matter for theologians and philosophers to ponder.

What we need to remember is this one simple fact. No society, nation, or people who order their lives around a service-to-self paradigm (world view) survive the test of time.

It is because the service-to-self paradigm obfuscates what is required for a viable survival paradigm for crossing the cusp with a desire for acquisition.

Conversely, viable paradigms can be found today in the folklore and traditions of ancient indigenous peoples, such as the Hopi and Aborigines. These belief systems seek harmony as opposed to acquisition and therefore follow a service-to-others approach.

In terms of surviving a pole shift, this very difference between these two groups is the point of truth from which the survival story begins.

The Noah Paradigm

Earlier, we used the deluge story of Sisuda and Hanok from *The Kolbrin Bible* as a prescient description for what lies ahead.

Yet, the Genesis story of Noah and the flood is what actually gives us the point of truth from which the survival story begins—Noah's own paradigm—and it shows us a proven, crossing-the-cusp survival paradigm:

> **Genesis 6:9**
>
> These are the records of the generations of Noah. Noah was a righteous man, blameless in his time; Noah walked with God.
>
> One cannot be anointed or blessed as being anointed and blameless. Nor will the last minute muttering of incantations and public donations of money achieve the desired result. It is how you chose to live your life.

In Noah's state of mind, he saw himself and his connection to all life on the planet. As a righteous man, he was upright and moral, and as a blameless man, his integrity was irreproachable. One cannot achieve such a state of harmony without a sense of oneness.

It is this connection of oneness wherein you continually seek harmony within yourself and all that it about you. Free from the noise of self-interested acquisition, you can see far beyond immediate desires.

Whether or not you like what you see, you nonetheless see it. As much as it may pain you, you are aware and you know that you must accept the burden of this knowledge, formulate a plan, and then act on that plan. Thus inspired, you are insulated from the taunts of the shortsighted.

Herein is the dilemma for a few, whether in government or privy to what governments know, about that which is to come.

The Noah Dilemma

Many come to know what is to happen through access to secret documents possessing far more data granularity than this book could ever hope to present. Likewise, a few find themselves in the Noah dilemma.

The Noah dilemma is one that forces people to make a brutal choice. Do they dismiss the future so as to keep their material advantages (which are considerable) in the present? Or, do they see themselves leaning over the edge of an abyss?

If they choose the abyss, they will lose themselves and their own sense of humanity. For most, the material gains far outweigh the price, but for a few, the price is too great.

These few know that going public means sacrificing their careers, income, security, health, and possibly their significant other as well. Yet, they pay.

While they seldom dwell on the cost, each will say the exact same words: Now I know how Noah felt.

It is easy to assume that they're railing against the mockery, derision, and coercion they must endure and to dismiss this as righteous indignation, a "just wait and see, you'll get yours" ego rant. However, only could Hollywood see it that way because only the whistleblowers understand the price of choosing to become "a righteous man."

A Righteous Man

When the story of Noah and the flood is taught, we feel sympathy for all mentioned in the story. For Noah, we feel a sense of commiseration and with those who drown beside the ark we feel a certain degree of compassion.

The visions that haunt whistleblowers enable them to empathize through the power of understanding the spiritual agony Noah would have felt as the deluge lifted the ark.

Outside he hears those who had arrogantly mocked and ridiculed him, pounding on the thick wooden hull of the ark and pleading for their lives. Only now are they ready to listen, but it is too late.

It is not that their urgent pleas fell on deaf ears. As a righteous man, Noah would no doubt have felt the deepest sense of sadness imaginable as his own compassion begged him to "open the doors."

Likewise, a calmer voice within would also remind him that this is a consequential moment of personal choices—the choices he and all the others made in anticipation of this cataclysm—both wise and unwise.

In this calmer voice is the imperative to "stay on mission," that being the survival of life on this planet. For this reason, Noah knew a bitter truth all too well. He knew these frightened and panicked people were completely unprepared for what was happening to them and their world. To open the doors to them would be to invite angry drama and the subsequent failure of the mission. Then, to save an arrogant few, all would be lost.

For the rest of us, the choice is best summed up with a simple question: Where will you be in 2012? That will largely decide where you will be when the pole shift happens. Will you serve the mission or fail yourself? These are the choices and their consequences.

Choices and Consequences

Preparing for global cataclysm is not a last-minute affair, nor are communal pleas for forgiveness and relief. Rather, preparation requires time and a deep abiding commitment.

It is why the wretched souls pounding on the stout timbers of the ark, if Noah had saved them, would become unmanageable risks to the ark's core mission of preserving life on the planet—the transcendent purpose in such dire times to which all others are subordinate.

Noah understood this, and in dimensions we can only imagine until we come into awareness. This is why whistleblowers forfeit the safety of their material cocoons for the poverty of integrity. Their thoughts inevitably are those of Noah in the darkness of many a lonely night.

In the stillness, they hear frenetic pounding upon the ark's hull, but in the context of what is to come. And so they feel a deep sadness in their soul, just as Noah certainly had. It is, indeed, a heavy, heavyhearted camaraderie.

The whistleblower comes to understand the most important asset for surviving a global cataclysm—the manner in which each of us chooses to walk with our Creator. To walk in the steps of Noah is to subscribe to the Noah paradigm—one that we know works. This is the reason why people all use another expression every chance they can: "You need to get your spiritual house in order."

This in turn raises a necessary question. If we are to walk this path, where is the starting point and how do we read road signs, those defining attributes of the Noah paradigm?

Defining Attributes of a Noah Paradigm

The defining attributes of a Noah paradigm are given in *The Kolbrin Bible* with a two-part prophecy. It comes from the Book of Manuscripts, and it speaks directly to these times:

> **MAN:3:10** In those days, men will have the Great Book before them; wisdom will be revealed; the few will be gathered for the stand; it is the hour of trial. The dauntless ones will survive; the stouthearted will not go down to destruction.

First we are told that we will have the *Great Book* before us. What remains of the *Great Book* are contained in the first part of *The Kolbrin Bible*. These passages were inscribed by the ancient Egyptians in the years following the Hebrew Exodus.

The second part tells us who will survive, namely, the dauntless (bold and intrepid) and the stouthearted (brave and resolute.) These are the defining attributes of a viable survival paradigm.

It is interesting to note that *Great Book* prophecies were first fulfilled in the early 1990s when the text was finally made public by a secret Scottish society. This same society had, for generations, continuously safeguarded these ancient texts since the Glastonbury Abbey was set ablaze in 1184 CE.

After several predictions were fulfilled, they revealed the texts. The one passage in the Book of Manuscripts where we also find the deluge and Planet X accounts speaks directly to the early 1990s with uncanny clarity, as is shown in the passage below:

> **MAN:3:8** A nation of soothsayers shall rise and fall, and their tongue shall be the speech learned. [Great Britain] A nation of lawgivers shall rule the Earth and pass away into nothingness. [League of Nations] One worship will pass into the four quarters of the Earth, talking peace and bringing war [Islam]. A nation of the seas will be greater than any other, but will be as an apple rotten at the core and will not endure [Soviet Union]. A nation of traders will destroy men with wonders and it shall have its day [America, First Iraq War]. Then shall the high strive with the low, the North with the South, the East with the West, and the light with the darkness [hegemonic globalization]. Men shall be divided by their races, and the children will be born as strangers among them [assisted reproductive technology]. Brother shall strive with brother and husband with wife [breakdown of the family]. Fathers will no longer instruct their sons, and the sons will be wayward [street gangs, terrorists]. Women will become the common property of men and will no longer be held in regard and respect [sex slave trade and plight of Afghan women].

To all things there is a time and this was that time. The time had come to reveal the texts. With that in mind, we return to the projection of the messages contained in the Avebury 2008 formation.

The projection begins in December 2012 and goes through 2014. It is a projection of an unbroken chain of catastrophic events, eventually resulting in a pole shift.

Flyby and Pole Shift 2012 to 2014

The central premise of Catastrophism is that evolution on this planet is driven by long periods of relative quiescence punctuated by very brief moments of global cataclysm.

Catastrophism in Deep Geological Time

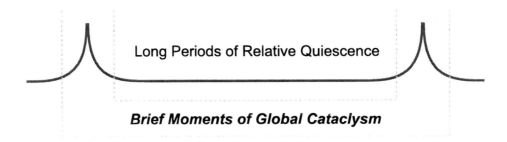

Flyby and Pole Shift — 2012 to 2014

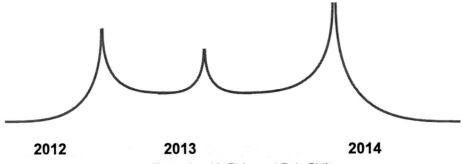

Illustration 43: Flyby and Pole Shift

That view of Catastrophism spans a vast period of geological time, and so in this projection, we must narrow the focus to a relatively small period of time.

This time line begins in December 2012, when the Avebury 2008 formation shows us that Planet X is entering the kill zone. This starting time heralds the first of the three phases described in this projection. The third and final phase in 2014 includes the pole shift through to the early days of the aftermath.

Phase One — 4th Qtr 2012 Through 2nd Qtr 2013

Phase one is when we need to go to ground to shelter ourselves from the horrible solar eruptions triggered by Planet X as it transits the kill zone. According to the Avebury 2008 formation, every living soul on the planet will know that life as we know is finally at an end.

First Phase

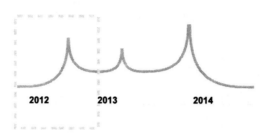

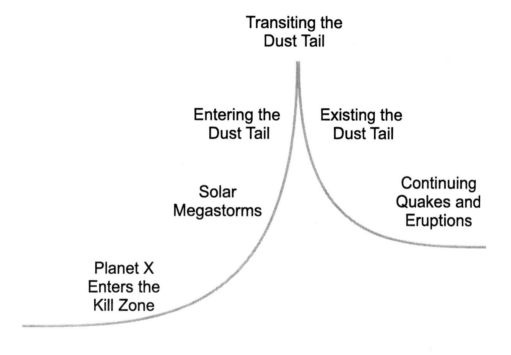

Illustration 44: First Phase 2012 - 2013

During this first phase, Earth will be battered by solar storms and pummeled with meteorite showers, bolides, and severe impact events. After our planet completes the first transit of the dust tail, earthquakes, aftershocks, and volcanic eruptions will continue to plague us.

Our planet can absorb such events, but only to an extent. Located in what astronomers call the Goldilocks Zone, or habitable zone (HZ), the Earth's orbit of the Sun puts us at just the right distance from the Sun to make life possible. This is a human-friendly point of equilibrium (PoE) as far as we're concerned, but all things can and do change.

Here is where the Second Law of Thermodynamics applies to the flyby. It tells us that when left to themselves, organized systems will become unstable and less organized as time advances.

It is likewise called the Law of Entropy because, in physics, entropy is the amount of disorder in a system. Ergo, when a system goes from a stable condition to an unstable condition, it experiences an increase in entropy.

In terms of a Planet X flyby, violent solar storms and impact events will fall upon us like relentless hammers of entropy and bring us to phase two of the flyby.

Phase Two — 2nd Qtr 2013 Through 1st Qtr 2014

Planet X will be somewhere between the orbit of Mars and the orbit of Jupiter at this time. Like a retreating comet, it will be visible to the naked eye as a diminishing threat, provided the skies are not fully obscured.

Second Phase

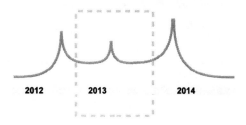

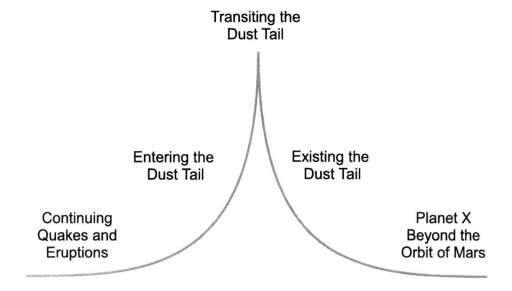

Illustration 45: Second Phase 2013 - 2014

Here is where Mother Shipton's warning of "false grandeur" suggests that there will still be limited visibility and why phase two will be especially unkind to the uninformed. Believing the worst has passed, they revel in their survival as they set about recovering the fragments of life as they knew it.

Midway through the second phase, even as Earth is dissipating the unusable energy accumulated in the first phase, we transit the dust tail of Planet X for a second time.

The catastrophic consequence of this second transit is that all these impact events re-energize the greater pool of unusable energy the planet must contain. This in turn prolongs the duration of the dissipation process beyond manageable limits.

For example, the 7.1 magnitude Loma Prieta earthquake that struck the San Francisco Bay Area on October 17, 1989, lasted for approximately 15 to 20 seconds. After the event, it was estimated that if the event had lasted roughly 10 seconds longer, every bridge and overpass in the Bay Area would have collapsed.

As with all things, there comes a time when there is no more time to dissipate unusable energy and the inevitable happens. Yet, many will be oblivious to this and, worse yet, cut off.

During this time, communication will be extremely limited, if available at all. Consequently, many who survive the ordeals of phase one can only guess about what is coming next and too many will guess unwisely.

The Downside of Hopeful Assumptions

By the time Planet X has entered the kill zone in phase one, the elites and their minions will be underground or off world Those left above ground, including the visible part of the government, will be nearly as blindsided as the general public.

Consequently, they will fall back on their training, which teaches them to restore order then wait for further instructions and relief. Unfortunately, there will be neither instructions nor relief.

They, too, will be sacrificed as "acceptable collateral damage" for the benefit of all humanity—that being the elites and their minions—but haven't we seen that movie once before?

Up to the Bottom

The second phase of this cataclysm will play itself out in many cases in a manner reminiscent of a visually powerful scene in the movie, *The Poseidon Adventure* (1972).

In the film, an undersea quake triggers a tidal wave. It slams into the great ship causing it to capsize. What was once the top of the vessel is now the bottom.

Having survived the Poseidon's death roll, survivors in the grand ballroom now face two survival options. Follow a no-nonsense reverend played by Gene Hackman on a perilous long-shot journey up through the bowels of the ship. Or, do as the ship purser says: Be calm and wait for rescue.

Thinking the worst has passed and hoping for rescue, most everyone chooses to follow the ship purser's instructions.

Those few who choose to follow the reverend start their escape journey upwards through the bowels of the ship. No sooner had they escaped the grand ballroom than the sea crushed in, drowning the ship purser and those who chose to followed him.

Beware the False Grandeur

With this Poseidon ship purser analogy in mind, two passages from Mother Shipton's prophecies offer vital warnings to those who live through the first phase of this global cataclysm:

> And when the dragon's tail is gone,
> Man forgets, and smiles, and carries on
> To apply himself - too late, too late
> For mankind has earned deserved fate.
>
> His masked smile - his false grandeur,
> Will serve the Gods their anger stir.
> And they will send the Dragon back
> To light the sky - his tail will crack
> Upon the earth and rend the earth
> And man shall flee, King, Lord, and serf.

In the movie, is the ship purser an evil man? No, just a man doing what he has been trained to do. The only problem was that the absence of specific training for situations such as this caused him to use what he thought would apply. This assumption was the false grandeur of their undoing.

The point here is that what is logical to first responders and leaders in a time of relative quiescence may be effective or dangerously counterproductive during a global cataclysm. Ergo, be not free and easy with your trust.

In Act I of Hamlet, William Shakespeare speaks great wisdom through the character of Polonius as he counsels his son Laertes for travel abroad: "This above all: to thine own self be true,/And it must follow, as the night the day,/Thou cans't not be false to any man" (ll.78-80).

A more pithy way of saying it is *trust, but verify,* the signature phrase of America's 40th president, *Ronald Reagan*. Regardless of your political persuasion, these three words will be life-saving words.

Repeat them each time you hear the siren call of false grandeur. They will help keep you focused on your own preparations and you'll need them, for the worst is yet to come.

Phase Three — 2nd Qtr 2014 Through 4th Qtr 2014

The third and final phase of this three-part global cataclysm will culminate with a pole shift as the unusable energy accumulated during the first two phases pushes our planet past its point of equilibrium, resulting in a pole shift—the ultimate releaser of pent-up, unusable energies.

Third Phase

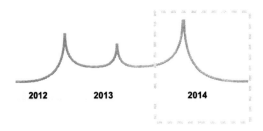

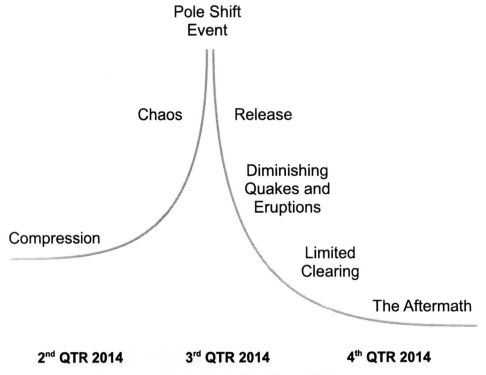

Illustration 46: Third Phase 2014

Assuming there is any visibility left in the skies, Planet X will be seen as it vanishes from naked-eye view. The beast will be returning to the frigid depths of space and, yet, the worst it will do will be like a terrible revenge—a dish best served cold and over-brimming with resonance.

The Resonance Effect

The instrument of Planet X's terrible revenge is something that could be described as the resonance effect.

Military leaders understand it quite well. It is why they always march their troops across a bridge with a route step, so their formations essentially stroll across the bridge.

This is because if they march their troops across the bridge with an organized cadence, the rhythmic thumping of the boots would resonate within the bridge itself.

If the column of soldiers is long enough, the unusable energy of that effect will undermine the bridge's structural integrity and cause it to fail, perhaps with troops on it. The same holds true in this case, where the unusable energies of the second phase resonate with what remains from the first.

Therefore, the three components of this resonance effect relevant to this discussion are: negative feedback, positive feedback and the catastrophic tipping point.

Negative Feedback

Negative feedback is when a system responds to a perturbation in a way that reduces its effect. A good example of negative feedback is noise-canceling headphones.

For years, they were expensive travel luxuries offered to frequent travelers on modern commercial jets. Thankfully, they are more affordable now though the technology remains essentially the same.

The positive feedback of intrusive cabin noises is neutralized or greatly diminished with negative feedback technology.

Positive Feedback

Positive feedback occurs when a system responds to a perturbation by increasing the magnitude of its effect. This term is often heard in conjunction with issues such as Earth changes and global weather changes.

For example, as global temperatures rise, the ice sheets melt. This exposes more dark surfaces which absorb the sun's warmth, whereas ice and snow reflect it away. Then a positive feedback loop starts where the more the ice sheet melts, the more heat is absorbed, which exacerbates the process and so on.

One could say that there are human equivalents as well. For example, when couples argue by taking turns jabbing each other under the belt, we know this behavior can turn a momentary disagreement into a relationship-ending confrontation. This is because each new jab reopens the wounds of previous jabs, which in turn begin to feed one upon the other.

Had the couple allowed time for their tempers to cool, negative feedback could have dissipated enough anger energy to reconcile their differences as opposed to pushing their relationship past its tipping point.

Catastrophic Tipping Point

A catastrophic tipping point is where the resonance of cumulative positive feedback events eventually undermines the stability of the system, resulting in a catastrophic failure.

For example, when a ship founders, it can take a relatively long time for the flooding to finally reach the tipping point where negative buoyancy takes over. However, once that threshold is crossed, the ship's fate is sealed and it begins to rapidly slip beneath the waves.

Had the ship's damage control teams managed to seal off enough leaks to keep the ship barely within positive buoyancy, the disaster could have been averted. This is because they would have broken the chain of failure, and catastrophes are the result of unbroken chains of failure.

In human terms, the process is somewhat like the bitter arguments of a quarrelsome couple, which can cause the relationship or marriage to fail completely. Like a foundering ship, it can take a long time for them to lose their positive buoyancy, but once they do, the end is inevitable.

Earth's Inevitable Tipping Point

In this scenario, the positive feedback begins in phase one with the solar duels between Planet X and the Sun. During this period, the Earth is bombarded with high levels of solar radiation.

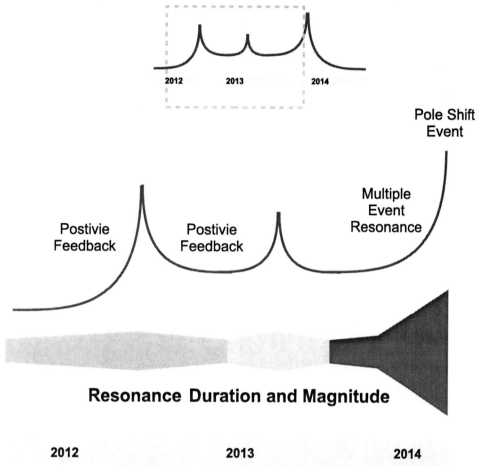

Illustration 47: Multiple Event Resonance

Of particular concern to us is that Earth's weakened magnetosphere will subject the surface of the planet to the heated plasma created by coronal mass ejections (CME) and at staggering levels.

All this unusable energy will go into the core of the Earth like firewood into an old cast-iron stove on which sits a full tea kettle. As you stoke the fire with more wood, the cumulative effect of increasing heat brings the kettle to a full boil.

In a similar way, phase two adds the energy of new impact events to the existing energy left over from the first stage, thereby setting in motion a sustained, positive feedback effect. Then, in phase three, we come to a full boil, so to speak.

The resonance of this sustained, positive feedback pushes the Earth beyond the tipping point. Once we cross that Rubicon of disorder, we're on our way to a pole shift.

Then, like a foundering ship at sea, once we lose positive buoyancy, we're going to see the ocean floor in a whole new way.

The Pole Shift

There is a long-standing debate among Planet X researchers about the possibility of this object passing close enough to the Earth that its mass alone could cause the pole shift via tidal gravitational forces.

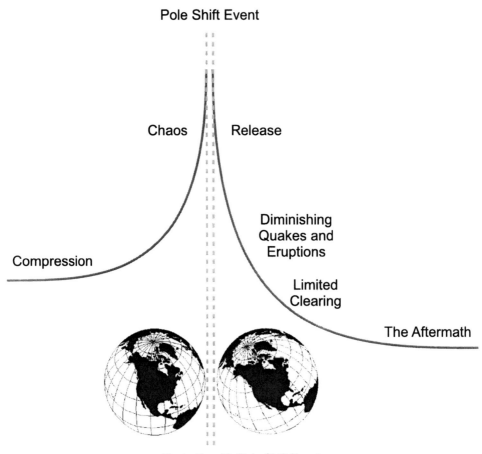

Illustration 48: Pole Shift Event

Assuming Planet X is big enough and comes near enough to the Earth for its mass to gain a gravitational lock on Earth's lithosphere, could it literally drag Earth around to a new orientation? Is it a gravitational tractor beam of sorts that serves as a convenient, single cause for a pole shift?

In that case, Earth is already in the grip of a gravitational tractor beam and it's coming from our own moon—the one object in our solar system that exerts the strongest tidal gravitational force on the Earth. Our moon.

If the Moon Disappeared

Were the moon to disappear, it is estimated that the surface of our planet could sink by some 18 inches, leaving the wondrous vitality of our oceans to stagnate.

However, more to the point, is what the crop circle makers have shown us in Avebury 2008 formation: Although Planet X will be close enough for tidal gravitational forces to come into play, this alone does not cause the pole shift.

Likewise, nothing in the ancient wisdom texts and folklore describing previous flybys supports a single-cause notion.

However, we can expect Planet X to exert tidal gravitational forces on the Earth as part of a larger, catastrophic one-two punch.

A One-Two Punch

The first punch will be the massive amounts of energy our planet absorbs as the result of solar eruptions and impact events. This is the positive-feedback resonance that triggers the pole shift.

The second punch will be the pull of Planet X's tidal gravitational forces upon the Earth. It will likely act as a pole shift direction mechanism.

Were the Earth thought of as a ship at sea, solar violence is what powers the vessel through the water. Planet X, on the other hand, will serve as a gravitational rudder that steers us through the shift to a new orientation.

While this simplistic explanation of how the Second Law of Thermodynamics applies to a pole shift gives the discussion an arm's-length feel, the process will be felt by everyone, and future generations will remember this time as "crossing the cusp."

It will be a time of cosmic disorder that brings about the single largest die-back event in the history of our species, and one that will be the most transformative as well.

The Greatest Single Cause of Death

The closer the Earth comes to the tipping point, the more the stress will be felt by all living species. Born of chaos, the period of the cusp is a time of flux and disorientation for all life on the planet. Those who survived phases one and two will now face the ultimate test of survival within the cusp between two ages, when the space/time continuum as we now experience it morphs into its next incarnation.

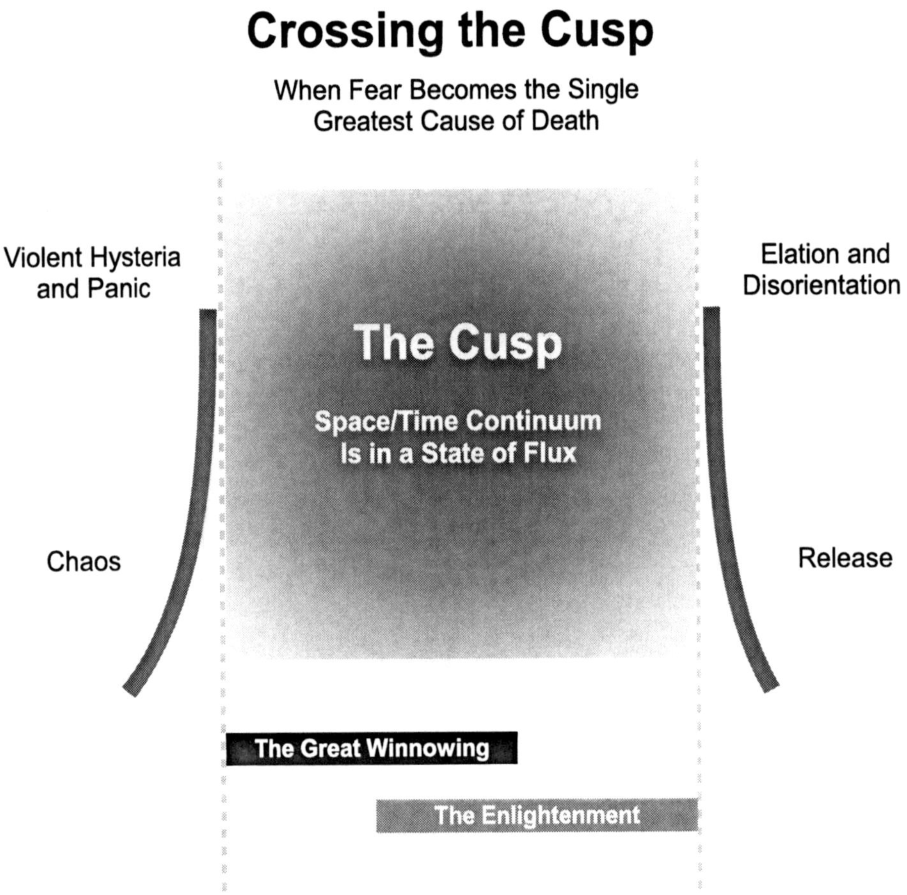

Crossing the Cusp
When Fear Becomes the Single Greatest Cause of Death

Violent Hysteria and Panic

Elation and Disorientation

The Cusp
Space/Time Continuum Is in a State of Flux

Chaos

Release

The Great Winnowing

The Enlightenment

Illustration 49: 05-the-cusp

Chaos Before the Tipping Point

Like the surface of our planet, we are mostly made of water. Ergo, that which perturbs Mother Earth can perturb us as well, and this will be a terrible testing of our metal.

As we approach the tipping point, signs of what is to come will be immediate and terrifying. Compasses will swing wildly, birds and animals will be disoriented, and weather patterns will become nightmarish.

Consequently, we can expect those who approach this event in a fearful state to commit unspeakable atrocities.

However, those who choose to approach this event from a basis of love (which begins with courage through integrity) will be ready to fend off, or in some way cope with, these atrocities.

The point of truth here is this: How you deal with your fear is what will likely determine your chances of successfully crossing the cusp.

Fear Is the Knife's Edge of Death

It was important to the creators and caretakers of *The Kolbrin Bible* that we living in these times understand what worked and what did not during previous flybys.

What doesn't work is fear. This is because fear is the knife's edge of death. Follow your fears and you'll inevitably stumble upon a blade.

In two passages from the Book of Manuscripts, the ancients share with us what they found to work, along with their heartfelt hopes and prayers for us:

> **MAN:3:10** In those days, men will have the Great Book before them; wisdom will be revealed; the few will be gathered for the stand; it is the hour of trial. The dauntless ones will survive; the stouthearted will not go down to destruction.
>
> **MAN:3:11** Great God of All Ages, alike to all, who sets the trials of man, be merciful to our children in the Days of Doom. Man must suffer to be great, but hasten not his progress unduly. In the great winnowing, be not too harsh on the lesser ones among men. Even the son of a thief has become your scribe.

In these two passages we are told that the "dauntless" and "stouthearted" shall survive "the great winnowing," which is the operative phrase here.

The Great Winnowing Within The Cusp

The three religions of the desert God, Judaism, Christianity, and Islam, await a judgment. What distinguishes them from the ancient Egyptians is that while they expect the judgment of a deity, the first-hand experience of the ancient Egyptians tells them otherwise.

The judgment will be a time when the natural laws of the universe separate or distinguish valuable humans from worthless ones. They call it "The Great Winnowing" and it accurately describes what we will experience during a pole shift, beginning the instant we first enter the cusp.

Bordered on two sides by a differential time/space continuum, the cusp will produce a powerful psychoactive effect upon all who enter.

Similar to that of Ayahuasca or LSD, for example, the psychoactive effect of The Great Winnowing will be unlike anything ever experienced by Westerners. This is because this experience will last for days and not hours. For those who enter in fear, the 1960s era term "bad trip" understates the experience they can expect to have.

A Cosmic Bad Trip

One outcome of experimentation with Ayahuasca has been the tragic deaths of tourists to the Amazon regions who try this powerful hallucinogenic alkaloid. The victims typically die as result of being poorly prepared for the experience by untrained Ayahuasca tour operators. A similar example includes the LSD bad trip complaints of Vietnam War-era hippies.

In these bad trips, tourists experiment with the drug without being prepared. This reckless approach often sets them up for a fearful response to the disorientating sensations and states they experience.

Unless they can control these fearful responses as the drug takes hold, the drug's effect will trigger a positive feedback loop where fears grow exponentially. Absent a counteractive, negative feedback effect, the fear compounds itself, resulting in a severe health crisis or, in some cases, death.

A fully prepared tourist, or one immersed in the experience with the help of a genuine guide, will be able avoid the slippery slope of fear. Many of these, if not all, will then experience enlightenment.

However, for the greater share of those who have survived up to this point, this is where their own fears will guide them to the tipping points of their own making. This will begin immediately, in the first handful of hours and days inside the cusp.

This is why the ancient Egyptians called it The Great Winnowing. It will be the worst imaginable "bad trip" for the fearful.

When the Great Winnowing Begins

The cusp will be remembered by future generations for many things, but one will stand out from the rest and it will be a number—an epitaph to all who perished inside the cusp. And generations from now, when the number is spoken, the response will be "never again." In the Book of Manuscripts from *The Kolbrin Bible*, we're told who will not live to utter those words. They are those who succumb first to the psychoactive effect of the great winnowing:

> `MAN:3:11` In the great winnowing, be not too harsh on the lesser ones among men. Even the son of a thief has become your scribe.

The morally weakest will be the most prone to fear. Consequently, they will also be the most likely to succumb early on to the cusp experience.

It is why the son of a thief beseeches the universe to be kind—a noble gesture, but the universe is not steered by human peculiarities. It is giving us as individuals and collectively as a species a brutally frank choice.

We can evolve and flourish or devolve and perish. Above all else, it is about choice. Regrettably, many will choose to suffer the fate of the weakest.

Fate of the Weakest

As the weakest first enter the cusp, The Great Winnowing will immediately begin to magnify their fears. The onset will be so rapid and severe that a positive feedback loop will geometrically magnify their fears until they literally die from fright.

Age, health, and circumstances will factor into who goes first and who goes last from among this group. However, all will meet the same end, so these factors are moot.

However, it needs to be understood that this will not be a prolonged or physically painful death. The weakest will just die from fright.

Like failing light bulbs, they may flicker a few times at first, but then comes the inevitable darkness. Following the weak will be the forceful.

The Forceful Fall Next

Those who through sheer force of will control their own fears and the fears of others will be the next to fall.

These people have the knowledge of fear and use it to manipulate and exploit others. Part of that ability is an insulating sense of detachment, and here is where the sheltered emptiness within them becomes their own tipping point.

They will see the weakest falling early on into the cusp and be unfazed, and their force of will is bound to be strong at first. Then their veneer of strength will fail them and leave them standing naked before their greatest fear: that there really are no clever ways to beat the system.

What the Forceful Fear

In evolutionary terms, passage through the cusp, from one side to the other, will be a passage through the membrane of destiny.

Imagine this membrane as though it were a kitchen colander filled to the brim with big plump raspberries. In this metaphor, humanity is the water and each raspberry a tiny drop within it.

As water droplets fall, the job of navigating the tricky terrain of a raspberry is manageable to a degree. However, the hard part is making it through one of the holes in the sides and bottom of colander, which all lead out to the other side of the cusp.

The weakest will be sure to run aground on the convoluted surfaces of the raspberries. Likewise, those with the force of will to go beyond them will cast indifferent glances backward at them, certain in the self-belief they'll find a way to cheat death with a clever solution.

Whether by duping someone else, discovering a secret passage, or whatever, they will make it. They will try with great vigor but eventually come to realize their worst fear. The self-made emptiness within them is where those who live in service to others find their path to survival. And they will not even know how to recognize it.

Granted, luck will favor a few, but most will sink into a futile sense of hopelessness and their greatest fear will exhaust them. Then The Great Winnowing will take them, as one would gaff an exhausted game fish, weary of the fight.

Will most elites and their minions meet this same end? Perhaps.

Elites and Their Minions

No matter how deep one burrows into the Earth to survive the radiation of an angry sun, one can never burrow deep enough to survive The Great Winnowing in the cusp.

Elites understand this phenomenon very clearly and they know it will pervade the entire planet during the shift. No matter where you are on the Earth, The Great Winnowing will reach you.

We can be certain that elites and their minions have factored this into their plans and some will go off world to escape The Great Winnowing, as unbelievable as that may sound.

However, the larger part of them will be heavily sedated or perhaps placed into drug-induced comas just prior to the onset of the pole shift event. Whatever they do, it will be a technological approach, with all that entails.

Another thing for which they are preparing with great vigor is their return. Not only have they decided the matter of human destiny after the pole shift, but they also intend to see it through.

That being said, one fact remains. They're working from educated guesses and self-interest, and as clever as they may appear to be, at the end of the day, they're not gods.

Nor are they immune from the choice the universe is giving us both as individuals and collectively as a species: to evolve or devolve.

In this evolutionary event, those destined to survive and evolve will find their destiny path within the cusp to enlightenment.

Having avoided the experience, elites and their minions will return to a world they will not recognize. Nor will it care to recognize them, leastwise, not as they would wish.

They will reappear exactly as they are today—the failed anachronisms of an unsustainable way of life.

Despite their cleverness and force, this is our chance as an enlightened species to see to it that they never find an apple from which to take a second bite.

Failing that and the elites prevail as before, humanity will devolve into a slave species for countless generations to come.

Then, only after our world has been brutally raped of its resources and beauty will we be given our freedom again—for whatever pitiful sliver of time remains.

For this reason, the transcendent issue is how we, the common folk of the world, choose to approach this coming global catastrophe.

If enough of us can replace our fears with great concerns and do our best to walk the path of Noah, something magical can and will happen. The coming tribulation, as terrible as it will be for us, will also be the birth pain of a new us as an enlightened stellar species.

The Enlightenment

Those who have entered the cusp in a state of love will initially be focused on avoiding confrontations with the fearful, and it will be necessary for their survival.

As the population of the weakest and forceful reaches a tipping point, the threat they present to those in a state of love will diminish along with their numbers, and this will be the harbinger of the enlightenment.

Human destiny is a book of many chapters. The present chapter is the sole domain of the elites and their minions and they intend to extend their control to the next, but not this time. Humanity's numbers are staggering and, within these vast numbers, the statistical push is finally able to work in humanity's favor.

It takes enough seeds to plant a thriving orchard and, this time, there will be more than enough. The universe has given us this last and best chance and Planet Earth is ours to use or to lose.

If enough of us come to realize it, which is the true purpose of this book, the stout-hearted, and not the elites, will write the next chapter in the story of human evolution

Today you too must choose, but before you do, know this. Only those worthy of holding this noble pen of evolution will be capable of crossing the cusp—and they will.

Part 2
Crossing the Cusp

The Kolbrin Bible
MAN:3:10 The dauntless ones will survive; the stouthearted will not go down to destruction.

6

Welcome to Awareness

One day, others will ask you the "who, what, when, where, why, and how" questions. This is why only one word truly describes the first part of this book—brutal. But, then again, which would you prefer? The vaccine or the virus? Such is the burden of this knowledge.

You'll have more of this knowledge than most, and they'll see it in your eyes. In those days, you will speak with them, often in one-on-one conversations. It may seem unimaginable now, but look inside. See that future time when you'll be answering the very kinds of questions that are now buzzing about in your mind.

A Future Time for You

With this mind, let us speak one-on-one and assume that we've just cleared the dinner table. While the teapot slowly comes to a boil, our conversation moves beyond the typical pleasantries of idle dinner chatter to matters of great concern.

Yes, in that future time, there will be no tea and cake. Just sad eyes staring at you with a deep hunger to understand the meaning of what has happened and, perhaps, catch a glimmer of what is yet to come. Do you see yourself accepting that responsibility? Can you? If you are truly to survive, you must.

As one person to another, let's put the cards on the table. To answer these questions, I'm going to turn the tables on you and ask you to visualize a time just following a catastrophic event.

If you have survived, it will be through a combination of good fortune and your own preparations. While percentages will vary, you will have enhanced your odds by first accepting the brutal blow of awareness. From that you pursue knowledge, which in turn becomes preparation. You will do it this way because acceptance takes less life energy than denial. In fact, it will, in time, increase your energy.

Most others will be ill-prepared for what has happened. After surviving the initial events, they will be stunned and highly vulnerable to ensuing calamities. Many will give up and die from despair. You'll see them, and you'll also see those with the will to endure.

They in turn will see you because they will be looking for someone like you, who possesses a clearheaded understanding of events. Though shaken, they will witness how you rebound most quickly of all, and with a presence of mind that keeps you intelligently focused on survival.

In a sea of hysterical reactions, you will stand out because your knowing will be the result of a natural, transformative process—on both the physical and spiritual levels. Taking the first step is simple, yet this is where the vast majority stumble. So, what is that simple first step?

Look Within Yourself

The key to your own survival is inside you. You were born with it, and the engrams will automatically kick in. For example, let's imagine a man who is angry that his lover carried an unplanned child to term. The instant he first holds his child in his hands, he turns from a disgruntled tyrant into a doting and loving father. With exceptions noted, the programming is there and it automatically kicks in. We would not be here today, the dominant species on the planet, were it not for this programming.

The same holds true for surviving the coming global cataclysm. The ability to deal with it is something we're all born with, and it's deep inside our genetic code, ever vigilant to a waking moment. Think of your waking moment like a manual fire alarm, the kind that says, "In case of emergency, break glass and pull fire alarm."

An Amazing Thing to Watch

In this case, your waking moment is that brutal jolt of awareness you received in part one of the book, which punched through the glass shield covering your fire alarm lever of consciousness. Now the question becomes, do you or do you not pull the lever?

That is a hard decision for most, because they sense that awareness leads to understanding things as they are. The consequence of that are feelings of anger, betrayal, and, worst of all, a hopeless sense that you've lost control over your life.

Consequently, many become paralyzed with fear of the immediate consequences without looking ahead beyond the initial upset. If they would, they'd see that the sooner they pull the lever, the sooner they will be back in control of their life. Frankly, it's an amazing thing to watch.

Having been researcher and author on topics of Catastrophism for more than a decade now, time and again, I've seen people go through this brutal introduction and find the strength within themselves to pull the lever. When they do, the outcome is invariably the same. They're off to the races like self-sufficient, confident Energizer bunnies. As a researcher, it is a most life-affirming process to witness. It truly is a testament to the power of awareness.

Once you are in awareness and seeking knowledge, this one cardinal rule will serve you well: The greatest truths are by and of necessity—simple.

A knowing can be composed of a number of many great truths, but only if they are bound together, without the need for obtuse exceptions and arbitrary rules. The dots must connect cleanly. If not, look to see if there are intentional obstructions, as here is where the evil slime of disinformation tends to percolate and grow.

The whole point of disinformation is to cause you to miss the point so that you draw erroneous or useless conclusions that only serve to exhaust your interest. Unlike propaganda which is a lie painted as truth, disinformation is always a mix of truth and lies.

It has just enough truth to earn your confidence and just enough lies to send you down one rabbit hole after another, until you throw up your hands and walk away. It is the favored tool of elites and their minions. Why do they employ it? To save themselves, protect their schemes, and further their empires.

This brings us back to the questions you have buzzing around in your head right now.

Assuming you survive, similar questions will be buzzing around in other people's minds, and they'll be looking to you for answers. So, let's start with the big one: Why weren't we warned earlier?

Why Weren't We Warned Earlier?

We are being gently warned to wake up to what is happening, but most people live their lives with a thumb permanently buried in the snooze button of their awareness alarm clocks, so to speak. In other words, how do you wake up so many who are working so hard to remain asleep?

What's lacking is the proper context, which happens to be the end of life as we know it—not as some form of religious judgment, but rather as a result of a cosmic disaster that cannot be nuked or negotiated with.

As a Catastrophisim researcher, I've found movies and cable television documentaries to be a rich and growing source of knowledge and warnings. Is this Hollywood acting on orders or simply doing what it does—measuring the pulse of public interest and satisfying it? I believe it to be principally the latter.

In recent years, we've seen a stream of catastrophe films, including *The Day After Tomorrow* (2004), *Children of Men* (2006), *Knowing* (2008), *The Road* (2008), *2012* (2009) and *The Book of Eli* (2010). Additionally, the History Channel, the Weather Channel, the Science Channel, and others present a steady stream of programs on prophecy and catastrophe in our times.

The transformation in cable TV documentaries has been tremendous. Throughout the late 20th century, prophecy and catastrophe programs were always presented as entertainment with lots of speculation and nay-saying to shield the producers with a supposedly "balanced" approach. The result was that most programs left the viewer with a feeling of being entertained and largely clueless.

Since the millennium, a trend has emerged. Many of these cable TV documentaries are now frank and straightforward in their approach. What viewers now see are serious attempts to present information in a constructive manner so they can draw intelligent conclusions on their own.

If you want to see a realistic post-historic, nuclear winter scenario, *The Road* (2008) starring Viggo Mortensen is the only authentic depiction to come out of Hollywood so far. It is hard to watch, but this makes it a truthful warning. All the other films are gentle with their warnings, many of which are buried under special-effects eye candy and predicable, formula-driven screenplays.

But there are hidden gems. For example, in the movie *2012*, director Roland Emmerich presents the eye candy we expect from Hollywood animation studios. However, if you watch the film a second or third time, you'll see a profound message buried underneath the glitz.

In *2012*, Emmerich shows audiences exactly how elites and their minions think and behave. Emmerich lays it out just the way it is, because in the process of creating this movie, he saw it as well.

Emmerich clearly sees what researchers in this field have long seen and known to be true —that the elites have known about what is coming for decades, if not centuries. Consequently, if you want to understand the elites and their minions, watch the film a second or third time.

The Green Revolution Sputters Out

What elites see is the end of an era often referred to as "The Green Revolution" as food production sputters to much lower levels. "Green Revolution" is a term coined for a period covering the 1950s and 1960s, which saw the industrialization of agriculture—a carryover from WWII, where the need to produce more food to support the war effort set the stage for the manner in which we now feed ourselves.

By 1984, these wartime goals had energized global production levels up by two-and-a-half times that of early wartime levels. The Green Revolution comprised a whole range of techniques and technologies; however, what powered it was the very same thing that powered the machines of war—fossil fuels.

Petroleum-based nitrogen fertilizers and pesticides are the principal drivers of industrialized agriculture. With aggressive application, these synthetics can triple or quadruple cereal crop yields. This type of fertilizer was regarded as the most important invention of the 20th century, more important than vaccines, spaceships, and antibiotics. How could this be?

Prior to WWII, farmers used organic methods with natural fertilizers, and the global population was just over two billion. Relying principally on family farms, this was largely a sustainable system. However, with the industrialization of agriculture, family farms languished.

Meanwhile, global population more than tripled within the span of a farmer's lifetime. Ergo, the trend lines of global population growth and crop yields per acre complement each other all the way up. However, this new approach to food production is not as self-sufficient as the less efficient, family farms of the prewar era. This is because agriculture today is just a subset of the greater oil energy industry.

What does that mean for America? At present, it takes 10 calories of petroleum energy to create one calorie of human nutrition. Elites understand that without enough oil to plant, fertilize, harvest, and transport crops, food production will dwindle. A bad turn for a world in which the human population is nearing the seven billion mark.

We could imagine this as a balloon within a balloon. The outer balloon is petroleum. The inner balloon is industrialized agriculture. As the greater oil balloon is squeezed by wars, shortages, and economic failures, the first responses will focus on absorbing the impacts as gracefully as possible.

Consequently, the smaller industrialized agriculture within the larger oil balloon will not feel the squeeze at first, but in time it will. Therefore, yields of vital food crops could drop by 50% or more. This is when everyone will really begin to feel the squeeze.

Here is where knowledge lost will come to haunt us. Most American know how to grow stuff...roses, herbs, small plots with cucumbers and tomatoes in the summer. How many, on the other hand, know how to grow enough stuff to feed themselves throughout the year—and this during a period of relatively benign weather?

Here is where the loss of so many family farms will cost us dearly. Consider this: At the outbreak of WWII, more than five million family farms in America could feed 133 million Americans and export food to the world. Family farms know how to feed themselves and others. Today, although America's population is 308 million, the number of family farms in America is one tenth what it was on December 7, 1941. Simply put, we're horribly vulnerable and in a way that was consciously and reckless engineered.

Starvation by Design

Attacks on oil production and transport have long been a consequence of regional conflicts in the Middle East; for example, Saddam Hussein ordered his forces to set fire to 700 oil wells in Kuwait during the First Gulf War. Until the 2010 BP oil spill, the Kuwait oil fires had the distinction of being the worst manmade ecological disaster in history.

It stands to reason that it is only a matter of time before these attacks become pervasive enough to seriously pinch the goo out of industrialized farming. Supply will become tenuous, as Islamist combatants work to disrupt oil supplies to a catastrophic level. Believing they can force the world to follow their agendas through oil starvation, they will seal their own fate and that of countless millions.

Given that America is the world's largest exporter of food, these disruptions will nonetheless spawn one export crisis after another, especially for those who depend on America. Weakened, America's system of industrial agriculture will not be the first casualty, nor will it be the last. In time, all available harvests will be reserved for domestic consumption so as to avoid, or postpone, open revolts by Americans.

For example, in August 2010, wildfires in Russia destroyed approximately 20% of the nation's wheat crop. Consequently, Prime Minister Vladimir Putin placed a temporary ban on grain exports, which sent a supply constriction shock wave through the futures trading markets.

Elites understand this is coming and partly fear it, but not for humane reasons. Well-fed people are much easier to control than starving people. With starvation, the ability of elites to control large populations will become stretched to the limit and, even with internment camps, will eventually fail. They know this and are mindful of history's lessons.

The Third Man on the Match

Today's elites have learned something that was well known in the trenches of WWI. Never be the third man to light your cigarette using the same lit match. Otherwise, a lurking sniper will put a bullet through your head, and before you finish lighting your cigarette. A French king and a Russian czar never got the message.

In modern history, the first time starvation caused monarchs and elites to lose power was the French revolution in 1789, which created the world's first democratic nation. It happened again in 1917 with the Russian revolution that created with world's first communist government. In both cases, monarchs and elites arrogantly ignored the hungry pleas of the starving masses and paid with their lives and fortunes.

That is why elites today, knowing they cannot feed the world during a tribulation, have a new strategy. They intend to prolong things as long as possible, so as to make good their escape to safety when the time comes to go to ground. Once they're gone, they'll be beyond reach. Meanwhile, starvation and other catastrophic forces will act in tandem to reduce human populations to much smaller and easier-to-control numbers.

While the common folks think in more noble terms (or fancy that they do), elites see it all as clear as day. For them, it is about maintaining control so as to preserve their position at the top of the food chain of life on the planet—now and after the cataclysm.

Elites know today that there simply will not be enough food produced on the planet to feed seven billion people, and the ones they're leaving behind are still mostly clueless. Their wake-up will come too.

It will be old and a bit scuffed, as it is the ghost of the Y2K millennium bug past. Beware, however, it will return with a terrible vengeance to kick out the last underpinnings of industrialized agriculture.

Truth or Self-Deluding Fantasy

Present crop yields not only depend on petroleum-based, nitrogen fertilizers, they also require a great deal of computer automation. Here is where industrialized agriculture has not one, but two addictions—oil and computers.

When you're sitting behind a plow horse, the global positioning satellite (GPS) overhead is really not going to be a big part of your world. However, if you're sitting in the air-conditioned cab of a massive, oil-powered harvester, GPS navigation is certainly a big part of your world, but not for long.

Back during the late 1990s, people were preparing for the Y2K millennium bug. This was anticipated fear that the computers of the 20th would fail to convert to 21st century dating, thereby causing a host of problems. The perception of this technology threat was polarized between the media and the public and those who really understood the threat.

The public's view of the Y2K bug was mostly shaped by silver and gold brokers, trading on hyped-up public fears. They all knew that corporations would prevent this disaster at great expense to themselves.

In the meantime, however, they could sell a lot of precious metals through this hyped-up scare, and the media went along for the ride. They enjoyed a lot of advertising revenue from Y2K panic headlines as well. For the brokers and media, Y2K was a free lunch, and the rest of the world was as dumb as a sack of hammers.

Those in the computer business who worked on Y2K projects knew different—folks like me—and we all understood that the public was largely being sold panic-for-profit disinformation.

If the media had really done their job, they would have explained to the public why information technologists across the globe were willing to spend a hefty percentage of their corporate information technology budgets on Y2K programming. Collectively, they could have financed a small war.

Why did they do this? Because they were able to understand the vulnerability of the national supply-chain management system. Helping them to do that was my job as well.

At the time, I was working as a freelance systems analyst and senior technical writer and was contracted by Hewlett-Packard to document what is clearly the Achilles' heel of the national supply-chain management system. In network terms, we know them as automated, rules-based messaging systems.

These messaging systems allow dissimilar computers running dissimilar software to share information in a predicable and useful way, and they're doing the same job today as they did back then. They provide an international digital grapevine, if you will, with zero tolerance for rumor and gossip.

Today, these systems are obviously no longer threatened by the Y2K bug; however, they are still as vulnerable to the solar activity predicted in the Avebury 2008 formation—a threat many times more dangerous than the millennium bug ever could have been. To help you understand the nature of the threat, let's eat some cornflakes.

Imagine you go to the grocery store today and purchase a box of cornflakes. You take your purchase to the checkout stand; the clerk scans the box; the transaction is relayed from the cash register to a back-office computer. From there, your purchase weaves its way through a long chain of intermediate computing systems culminating in the computer at the cereal manufacturing company.

All of this sharing of information about your purchase today happens with zero human involvement with only one exception, the clerk who dragged your cereal box across the scanner. Everything else is accomplished automatically using a predefined rules-based system.

Consequently, by the time you enjoy your first bowl of cornflakes tomorrow morning, the manufacturer's computer system will have automatically incremented production by one box while you were sleeping so as to replace your purchase.

For you, the consumer, here is your SPOT (single point of truth), from which everything makes sense. Frankly stated, your SPOT is a simple question: What does everyone do about that one-box-wide, empty niche on the grocery store shelf?

Position, Position, Position

When it comes to your local grocery store, the positions coveted by all manufacturers and producers is going to be eye level on the shelves, aisle end-caps, center islands, and checkout displays. We're far more likely to purchase products when we see them in these locations than in less favorable locations.

This is why manufacturers pull out all the stops to lock in prime locations and why they need to keep a continuous supply of product going to that store to maintain control of these shelf positions.

If their shelf position goes empty, the store will use another brand to fill the space. Consequently, they might have to go through an expensive process of pushing the squatters (as they would see them) out of coveted shelf placements.

Before the days of computers, grocery store shelves were not as tall or as densely packed. This was partly due to the fact that the nation used a manual supply-chain process at that time, subject to human error. Consequently, manufacturers needed to keep a 90-day supply of product in the supply chain, via jobbers, warehouses, distributors, and the like, between their manufacturing plants and the local grocery stores.

Financing this required vast amounts of capital. However, with computerized supply-chain management systems, the burden of maintaining a 90-day inventory needed for manual systems was reduced to just 90 hours. And connecting the dots of this automated 90-hour system are rules-based messaging systems.

The result is that our stores have become densely stocked because between them and the manufacturing plants, there is a small, thin, automated thread of supply. Although thin and stretched supply lines offer great savings to just-in-time manufacturers, they also carry the same risks of what doomed two invasions of Russia, by Napoleon and Hitler.

In both cases, the Russians knew that their enemy's supply lines would become stretched and vulnerable to disruptive, pinpoint attacks. During the winter, nature worked with the Russians, who successfully disrupted these stretched supply lines with catastrophic results for the invaders. What business managers understood about the Y2K millennium bug was our dependence on machines. They felt the stretch and did something about it before the first snows.

The need to maintain legions of workers and staffers to run a manual 90-day system was dramatically reduced with computerization. Like the loss of family farmers to mechanized agriculture, reducing the supply inventory from 90 days to just 90 hours also meant abandoning a population of workers who were capable of running a manual supply-chain system.

This is one reason why the threat of the Y2K bug sent such a chill through information technologists and decision makers. They realized that if enough of the antiquated, but critical nodes in national and international systems succumbed to Y2K, the whole system would be disrupted or possibly collapse.

One way to think of the international supply-chain management system is to imagine a dinner table piled high with tasty dishes. The computers and their automated business rules serve as the legs of the table. To create a crisis, all you need to do is kick one of the four legs from under the table. The result is that the dinner slides off onto the floor, and that is where everyone eats for weeks, if not months, until the system is restored.

This is why information technology decision makers spent such huge sums on Y2K preparation. They understood that the very survival of their own companies was at stake. Was this information available to the media? Yes, if they had bothered to look but, on the other hand, lurid stories about airplanes falling out of the sky and exploding nuclear reactors sell.

Most regrettable was that those in government and industry who kept this Titanic from hitting the iceberg were never given credit for averting a crisis—one that could have likely plunged America and the world into a global recession or depression. Such was the risk, and only a few knew it.

After Y2K, the media again were irresponsible, and pundits demanded great punishments for what they believed was a huge and expensive hoax. Of course, having a handy scapegoat to blame after you've lined your pockets makes good business sense.

Likewise, members of Congress sought to cash in on this new twist in the story and humiliated the real heroes of the day in mean-spirited hearings. So Congress, gold and silver brokers, and the media chose to suppress the truth so as to line their own pockets.

This brings us full circle back to the question: Why weren't we warned earlier?

Such a question presupposes that elites who control the government and media will subscribe to a higher standard of truth for something as big as an imminent pole shift.

Ask if the government lies, and only someone from the government will tell you no, and that someone has been told what to say. If it weren't a common belief, it wouldn't be an essential plot element in the movies.

So, be honest. How many people (you included) can you count on your fingers who actually believe without any reservation whatsoever that our government wouldn't lie about something this big?

Do another count. How many people (you included) can you count on your fingers who actually believe without any reservation whatsoever that the media wouldn't lie about something this big?

If you're holding up any fingers, ask yourself this one last question: Upon what is this presumption based—fact, blind faith, or self-deluding fantasy?

Therefore, before you ask why we weren't warned earlier, you must first determine if you're actually ready for the truth. Once you are, your life becomes less complicated.

If you're ready, it's time for the next question: Will the ghost of Y2K past return in 2012 with a terrible vengeance? Regrettably, yes, but not in the way you would first imagine.

The Ghost of Y2K Past

Our international supply-chain management computer systems are obviously invulnerable to another Y2K bug at this time, but their greatest vulnerability has never been properly addressed. This is a particular danger because the Avebury 2008 formation predicts horrific solar storms starting in late 2012 or early 2013.

Government, military, and nuclear research computers are housed in radiation-hardened processing centers, whereas civilian and business computer systems generally are not. Civilian systems are typically situated above ground and depend on local power grids.

While the buildings in which they're housed are often designed to handle earthquakes, floods, and similar disasters, very few are capable of surviving the EMP (electromagnetic pulse) of a terrorist EMP bomb or radiation from a massive solar storm as it strikes the Earth.

These kinds of threats will incapacitate the vast majority of these systems. This is when the opulence of the 90-day supply of products in our local grocery stores reveals a fundamental weakness in how the nation is fed. The result will be chaos, like a slow motion train wreck. This points to the vulnerability of our railroads, as well.

Without railroads, moving harvests from the fields to producers and manufacturers, and then moving packaged products out along a thin supply line to the local grocery store, becomes another component of the crisis.

Like modern supply-chain management systems, America's railroads have replaced manual systems with computerized switch controls, crossing warning systems, and train-detection and similar systems.

The result is that our railroads are just as vulnerable to a solar storm as any other computerized system.

As the automation breaks down, we'll smell perishables rotting in freight cars, pushed onto seldom used sidings as our store shelves are stripped bare. Therefore, anyone who thinks that a "usual" supply of food in the pantry will be adequate for most circumstances is a candidate for starvation.

Here you have a very simple choice. While good supplies of basic food stocks and relatively low prices exist, set aside enough to survive these contingencies—at least three months' worth of food.

Of course, there are always those who like to say, "I'll cross that bridge when I come to it." If that is your strategy, that is your strategy. Just be mindful of the toll because everyone who thinks this way will be fighting over whatever is left of the 90-hour food supply at the local markets. Also, forget about collecting discount coupons. Nobody will be taking them.

When shelves are bare and stay that way, people will understand that life as we now know it has come to an end. Not because they want to, but because denial will no longer work. Nonetheless, the vast majority of those who do survive the initial events will be needy and hungry. They'll also be angry and prone to pointing fingers and making demands for entitlements.

Now you have the full answer to the first and second questions. No doubt you have plenty more buzzing about in your mind, but instead of looking at what you do not have, take stock of what you do have. You are now in possession of this knowledge, so you can help yourself and others to survive.

After the first catastrophe, survivors will ask you: "Why weren't we warned earlier?" You can give them a long story, such as the one you've just read here, or you can give them the short answer.

We built an unsustainable world. If it hadn't been this, it would have been something else. Either way, it was inevitable so the real question now is whether we can build a sustainable world. If so, who will do it? However, before you speculate on that, you need to understand that elites do not want you to ask this question. They want you to be indifferent or confused. Either way, you'll go off into the weeds.

The Meek Shall Inherit the Earth

When it comes to employing disinformation, elites and their minions understand the need to manage how people in the mainstream see themselves and the world around us.

A mantra familiar to all researchers in this field is the disinformation blow-off. People say, "Oh yes, it exists. So what? Move on to something useful." It is like arriving in Los Angeles with just enough time to get to the meeting and the only map in your briefcase is for San Francisco. In other words, being in the same time zone doesn't show you the true lay of the land.

Mea culpa, I'm guilty of saying this mantra, too, and for the better part of my life. Then in 2008, after publishing articles on Yowusa.com documenting the NibiruShock2012 YouTube.com disclosure videos, I became the ground zero of an intense and mean-spirited disinformation campaign.

Fortunately, I had help in understanding it and how to deal with it, and I found this knowledge to be my best defense. In time, I came to realize that by countering the disinformation, the people slinging it at you still win. This is because the whole point of disinformation is to misdirect you, to distract you from finding a great truth by any means possible.

That is when I flipped my strategy altogether. Instead of fighting them, I started following them because disinformationalists have resources available to them that I could only dream of. Best of all, they are so easy to profile, if you know what to look for. Once you do, they are as easy to spot as gaily painted buzzards flying overhead.

When I see these buzzards start to circle, there is bound to be sweet meat beneath them, so get there before they pick it to pieces. As a result of this change of tactics, I've found hidden clues and honest sources I would have otherwise overlooked.

In terms of their outlook on life, disinformationalists are not meek people. They are more like petty, obnoxious retail clerks, and their methods are predicable and often clumsy. What makes disinformation work for them is that they can create enough of it to cause real harm. They also have a susceptible public to manipulate.

They're past masters of the game because people in general have no idea of how subtle and pervasive this disinformation really is. A classic example of subtle programming is how everyone defines the "meek" today.

If you look for a definition of the word "meek" in a modern dictionary, you'll see it described with terms such as compliant, cowed submissiveness, docile, spineless, spiritless, and tame. This is a complete contradiction to how it is used in the Holy Bible, where it is applied in a consistent manner:

- **Psalms 25:9:** The MEEK will he guide in judgment: and the meek will he teach his way.

- **Psalms 37:11:** But the MEEK shall inherit the earth; and shall delight themselves in the abundance of peace.

- **Matthew 5:5:** Blessed are the MEEK: for they shall inherit the earth.

Some dictionaries will tell you that this term dates back to the late 12th century and that it originally meant gentle, courteous, and kind. What's wrong with this picture?

To answer that question, look up these definitions for yourself so that you see with your own eyes. The act of physically seeing will give you a better grasp of the subtleties.

Start by creating a comparison table like the one below to connect the original meanings to any logical, present-day counterpart. No guessing allowed. You must have a solid, indisputable explanation of why the world "courteous," for example, is synonymous with "spineless."

Original Meaning of Meek 12th Century	Current Meaning of Meek 21st Century
Courteous Gentle Kind	Compliant Cowed submissiveness Docile Spineless Spiritless Tame

As you connect the original meanings with the current definitions, ask yourself the following questions:

1. How long have you understood the current definition and how often do you use it?

2. How long have you known about the original meaning and how often do you use it?

3. If you were an elite, which definition (obsolete or current) would best serve your purposes?

Consider this: If modern dictionaries are correct, then the people who inherit the Earth will be spineless and spiritless. Granted, elites are not likely interested in authoring or publishing dictionaries, but rather in making sure they serve their aims, as needed.

That is precisely what is happening here, whether you wish to believe it or not, because the elites have long known something that you've also known but never really understood. That is, they understand why the meek inherit the Earth, which has absolutely nothing to do with religion. Rather, it happens in spite of religion.

Hope vs. The Elites

One way of looking at how the elites program those they seek to control is to imagine that you're a young teenage girl and that you've been abducted while walking home from school one day.

Your abductor locks you in a small basement room and then tells you that everyone thinks you've run away from home; then he has his way with you. Over a period of days and multiple rapes, he probes your doubts and worries and plays upon them so as to reinforce his message that you are isolated and his to do with as he pleases.

However, the actual truth is that half the town is beating the bushes searching for you, hoping for any fragment of a clue to appear. Your abductor and rapist knows this, but the bluff favors him as long as you remain hopeless. However, if you refuse to give up hope and do all that you can to call for help, no matter how extreme the risk, you are calling your rapist's bluff and his control is no longer certain.

Nor does your refusal to give up hope and to feel isolated guarantee your rescue, but compared with the resignation of hopelessness, any chance to survive and be reunited with your loved ones, no matter how desperate or faint the hope might be, is far better than none at all.

The point here is that elites can only cheat the meek of their destiny when those they seek to control allow themselves to feel isolated and hopeless. Never, never, never give up hope that the meek will prevail! Never! Never! Never!

Once you accept this truth, there is only one relevant question and it does not deal with the past or pointing fingers. It is: What can I do today to help the meek inherit the Earth? When should you ask such a question of yourself? Now, today, and every day for the rest of your life.

7

How the Meek Prevail

In the coming tribulation, two important groups will play a special role in securing a positive outcome for the evolution of humankind: lightworkers and lifeworkers. Though they serve very different roles, their missions are equally necessary:

- **Lightworkers** come from diverse backgrounds (i.e., spiritual teachers, hospice caregivers, psychics) but all share the common goal of encouraging spirituality through sharing light, love, and harmony. They are not to be confused with fortune tellers, who are performers by nature, and in their forward-looking mission lightworkers serve as liaisons between this life and the next.

- **Lifeworkers** serve missions of life and death. This existential mission pivots between humanity at its best and at its worst. Lifeworkers put their lives on the line to protect others and come from a diverse range of backgrounds to include military, law enforcement, fire fighters, first responders, and critical care givers.

In the coming tribulation, a mutually supportive bond between these two diverse groups will enable them to closely work together, so that great things can be accomplished. This is not a far-fetched idea as there is supportive history.

Police departments, especially local enforcement, sometimes use psychic criminologists to uncover new clues to re-energize a stalled investigation. If the psychic can provide a useful clue, the investigator can move forward with either a direct lead to the criminal or a partial clue as to who the criminal could be.

There are no absolutes, yet there is a history of success where psychics have in some way helped investigators solve crimes. While this chapter is intended to be of benefit to all, it will speak directly to lifeworkers.

Now a moment of honesty if you will. As with everything else in life, there are those who remain true to their noble aims and those who take the easy money.

Over the years it has been my pleasure and great honor to make the acquaintance of many noble lifeworkers with a genuine interest in this topic. Good men and women who strap it on each day and put it on the line. As I write this chapter, these good souls and those like them are in my mind.

So, you noble lightworkers, please put on your psychic criminology hats and begin looking for dots, especially the ones that connect. If not now, maybe sometime in the future. In the meantime, just keep looking for dots and perhaps a clue will materialize to connect a few of these dots for you.

Lightworkers, do not feel left out, for this chapter is just as important to you as it is to lifeworkers. This is because it will help you better understand your future counterparts. As unimaginable as this may sound now, imagine it in terms of the future by seeing yourself in a local hardware store.

You'll be looking for a really strong adhesive and the clerk will suggests epoxy glue. In the packet are two tubes labeled "resin" and "hardener." The clerk will explain that once you mix the contents of both tubes, you will you have a really strong adhesive with which to work.

In essence, you and the lifeworkers are like the resin and the hardener in this analogy, and the coming tribulation will be what mixes you together. May the bond be strong and lasting.

For everyone else, here is where you come to learn how the meek actually do inherit the Earth, which begins with seeing how the concepts of entropy and harmony play out on a human scale.

Entropy

In Chapter 5, The Great Winnowing, we saw how the Second Law of Thermodynamics describes the amount of disorder in a system. There is also a human variant, where entropy represents a doctrine of inevitable social decline and degeneration. In the coming years, we will see both, and you will be on the front lines, which means you'll see it first—well before most everyone else.

As a lifeworker, perhaps you're already seeing harbingers of things to come. They may be the result of a disaster training program or something as simple as your reliable Toughbook laptop being replaced with a huge bulky laptop, radiation hardened to military specifications. Something big is in the wind, and this wind is blowing our world toward entropy—inevitable social decline and degeneration.

Throughout history, empires have followed the same path. They are born in a matter of a few generations and rise tall and mighty. Then, over time, they inevitably corrode from within as a selfish few manipulate the empire to draw much of the wealth and control to themselves at the expense of the many. The better they become at it, the more rapacious is their pace, as there is never enough to please the few.

Thus weakened from within, the inner core of the empire begins to rot like an apple at the core. Hence, the empire fails catastrophically from an event it likely could have survived during its prime. Then the process is repeated all over again, with a new cast of characters at the helm, all of the same ilk as those who created the first crisis. The faces change and the process repeats itself.

This time will be different. This time, we can have that better world that we dream about and you can play a vital role in making it happen. This is not a pipe dream, like the proverbial carrot dangled in front of a weary horse to urge him on. This time, that better world is truly possible.

What is your part to play? It is the same for everyone. Do what you do for love and always keep the faith. Do that and a fulfilling sense of harmony will be your reward.

Harmony

The process of The Great Winnowing is nature's way of selection and it favors the meek. This is because The Great Winnowing favors those in a state of order (harmony / love) as opposed to those in disorder (entropy / fear). If we return to the original and current meanings of the word "meek," a profile begins to emerge.

Original Meaning of Meek **12th Century**	**Current Meaning of Meek** **21st Century**
◢ Courteous ◢ Gentle ◢ Kind	◢ Compliant ◢ Cowed submissiveness ◢ Docile ◢ Spineless ◢ Spiritless ◢ Tame

When the word "meek" was first used, it described people who naturally favor harmonious relations, peace, amity, and friendship—all of which are used to describe the word "harmony." Therefore, the current definitions of meek are unnatural aberrations—or, in other words, disinformation.

This brings us to the critical point of the profile, where lifeworkers play a pivotal role. To understand that role, we must contrast fear and love.

Fear and Love

We all come from different walks of life, beliefs, and points of view. Nonetheless, all of our lives are bordered at the extremes by two simple absolutes: the human emotions of fear and love. All other emotions are but shaded variants of one or the other.

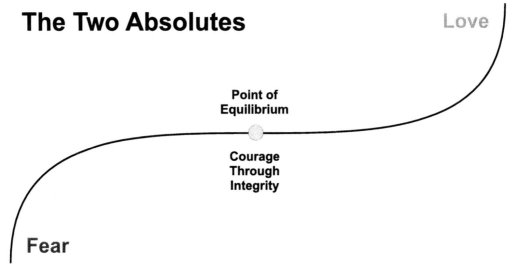

Illustration 50: The Two Absolutes

Imagine going to the paint store to order paint colored for one of the rooms in your house. The clerk shows you a vast array of colors to select from. Yet, no matter what color you choose the eventual choice will be closer to white than black, or vice versa. Such is the grand palette of human emotion.

However, in this case, we only need three, small palettes of four attributes each to understand the three principal profiles of those who will experience the tribulation.

Fear Profile (Entropy) *Inclined to...*	Point of Equilibrium *Courage Through Integrity*	Love Profile (Harmony) *Inclined to...*
Anger	Acceptance	Enlightenment
Apathy	Neutrality	Joy
Desire	Reason	Oneness
Pride	Willingness	Peace

Noble lifeworkers will be drawn to the Point of Equilibrium (PoE), and this is exactly where they need to be during the tribulation for this is what will get them through the times to come. Also keep in mind that the PoE is not at a point equally distant from fear and love. Rather, it is just a smidge in the direction of love, just like one of those pastel shades on a paint store palette.

For you noble lifeworkers, what gets you to this point will be your own courage through integrity because as you make it through this physical ordeal, you will see and experience things on a spiritual level, too.

Stay neutral, accept the knowledge, and set it aside for later contemplation. After the dust settles, the lightworkers will help you explore your experiences while crossing the cusp. The meaning of those experiences is for each to decide on his or her own.

Please note, I am not a religious man per se, nor am I a Pollyanna-ish New Ager. My only agenda here is survival knowledge, though you no doubt will recognize elements of different faiths or philosophies. It's all crumbs, and as I always like to say, "In this business you find the crumbs where they lay, whether they be on the table or the floor. Collect every morsel and savor it."

Also note that present Western belief systems all have a common weakness. None has ever proved capable of surviving a global cataclysm. Whatever your belief system may be, you must be honest with yourself about that. However, you don't need to throw the baby out with the bathwater.

Ergo, the views of spirituality put forth in this book are not intended to be theological or philosophical. It is about survival. Pure, plain, and simple and the view I now put forward is the result of a decade of my own research and contemplation on the topic. The concepts used here address the goal of successfully crossing the cusp and none other. However, what you believe is for you and you alone to decide.

To this end, I've created a minimalist paradigm for surviving a global cataclysm. The cornerstone concept of this paradigm is something I call the Natural Flow of Consciousness.

Natural Flow of Consciousness

Most folks, including those you are charged to serve, go about their daily business largely unaware of the world around them. They are focused on the short-term goals of the day: going somewhere, meeting someone, or doing something.

When you drive down the same streets, you see a very different world, comprised of good and bad alike. You know to look for harbingers of situations that could become hurtful, violent, or destructive.

For those of you who are truly honest lifeworkers, you're already at the PoE, so it is easier for you to expand your view beyond the physical reality of the world to one that reaches beyond the limitations of human imagination. To do this, start with a familiar concept.

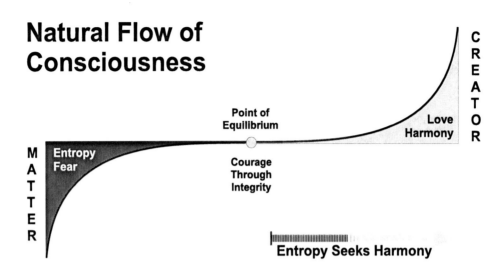

Illustration 51: Flow of Consciousness

People come and go all the time. It is our nature. On a spiritual level, the same holds true, but on a cosmic scale. Within us is a spirit; it existed before us, makes us who we are today, and will continue on long after the carbon used to make our bodies is reused by the cosmos to create new celestial events. Whether here or on the other side, your spirit is always on the go —always a work in process.

In a survival sense, the imperative question is: Which direction are you going? With or against the Natural Flow of Consciousness? One direction leads to life, the other to death.

In the physical world, entropy describes how ordered systems lose their equilibrium and fail. Ergo, the natural flow of the physical world is from an ordered state to that of a disordered state of entropy and failure. Lifeworkers see this every day. For the benefit of everyone reading this chapter, let's use a child abduction to illustrate the PoE concept.

On one side are parents who proudly behold their promising young daughter with the hope of good things to come—love, family, and excellence through achievement. On the other hand, pimps and slavers see this same child as a prostitution and pornography opportunity —one that can be easily seduced in many cases and then later controlled through violence and drug addiction.

Consequently, an innocent child who could be a successful member of her community is exploited as a wasted girl selling her body for the next fix. Lifeworkers see this needless waste of potential each day and often are the bearers of bad news to those least perceptive of the world as it operates.

This is a dark and dismal example, but take heart. In the spiritual world, the process is reversed. Entropy (disorder) naturally seeks harmony (order). It is a natural process bordered by inanimate matter on one side and the Creator (or however you choose to address the Creator) on the other. This world will be harsh for pimps, slavers, and all those whose lives are ruled by the fear they create for others as well as themselves.

This is why the PoE is one of personal choice. It is part of what is to come and people making choices about the present as well as the future. There is old expression, "We each die as we have lived." This wisdom is a pithy description for what will happen during The Great Winnowing.

Lifeworkers and The Great Winnowing

Noble lifeworkers can see in both directions from their PoE. To one side are those who define their lives with a fear-based, dog-eat-dog paradigm. Then there are those they prey upon, who are inclined to follow the natural flow of consciousness through love toward a closer connection with the Creator.

When humanity begins crossing the cusp, going with the natural flow of consciousness toward love is what will lead them through and beyond The Great Winnowing. Then the better part of their crossing the cusp experience will be that of enlightenment from beyond the cusp and into the aftermath, or the "backside" as I like to call it.

With my "Cut to the Chase" radio shows, I always close by saying, "This is Marshall and I'll catch you on the backside." Not in the CB radio sense, but where the backside represents the final moment of release from grips of this tribulation to come. For all of us, making it to the backside is the goal.

With or without the help of lifeworkers, the meek will make the crossing in good number. However, many more will make it to the backside with the active help of noble lifeworkers. Of these, the most important will be the innocents, our precious children. Within them is the hope of humanity.

Therefore, in a spiritual sense, you lifeworkers represent a vital, thin blue line in the coming tribulation. In addition to ensuring your own survival, your mission is to help those capable of crossing the cusp to do so in as safe a manner as possible.

Therefore, here is what lifeworkers need to look for, what they need to do, and when they need to do it.

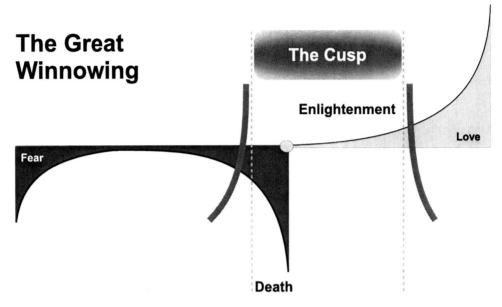

Illustration 52: The Great Winnowing

In Chapter 5, The Great Winnowing, we saw the cumulative effect of the solar storms and transits through the dust tail of Planet X. This will be a time of great stress for people and their governments and institutions, as it will magnify their own natural proclivities.

If fear (entropy / disorder) defines the person, then all of its permutations such as hate, greed, exploitation, and vengeance, will be exponentially heightened.

Conversely, if love (harmony / order) defines a person, that person's inclination will be to follow the natural order of consciousness; his or her proclivities will likewise be heightened. This is not to say these people will be perfect saints. Rather, it will be the tally at the end of the day that decides the matter, so to speak.

From 2013 on, we will see the ongoing implosions of world bodies, governments, monetary systems, extraditing systems, and institutions. Over time, the common people will become indifferent to these failing systems as the focus of life shifts back to a local level. During this time, some governments will fail outright and others, such as our own, will bifurcate into two variants: the more powerful below-ground government and the less powerful above-ground government.

Regardless of the promises made to those who left topside to manage things, few if any of the promises will be kept. Once the hope of support wanes, those left topside will come to understand that they are what they've always been—expendable.

As for the underground governments, as clever as they are, they'll still be dealing with a reality well understood by all military planners. Nothing ever goes exactly according to plan, that is, except in the movies.

In the days and weeks leading up to the actual pole shift event, anticipation of the event will compound the psychological effects. What everyone will experience is, in a matter of speaking, a powerful emotional accelerant, not much different than those used by arsonists to speed up the development of a fire.

The Accelerant Effect

As the pole shift event horizon comes close, the accelerant effect will become logarithmic in nature. Those in fear will naturally resist the natural flow of consciousness and this will be like throwing gasoline on a raging fire—first in cups, then in buckets, and finally in barrels in a monsoon-like deluge.

Consequently, those rooted in fear will seldom, if ever, survive The Great Winnowing. Their fears will be distorted by the psychotropic effect of the cusp, with an accelerant-like result. The processing will then feed upon itself, causing their fears to magnify within them. Once this process begins, it is uncontrollable and, eventually, their own fright switches off their lives, much like the last flicker of a dying light bulb.

This will be a hard time for those destined to cross the cusp, for they will hear what Noah heard inside the Ark: the pounding of fists on the hull and frightened pleas for help. Then shall they know what Noah knew: These fearful souls are beyond saving and attempting to save them would only place the greater goal of the mission in jeopardy. That goal is survival of life on this planet.

Therefore, lifeworkers must stay especially focused on those who can survive The Great Winnowing—the meek of the Earth. For them, the accelerant-like effects of the cusp will have the opposite result to those who live and create fear.

This is because an accelerant can also bring an organism back to a state of homeostasis, which describes a stable, constant condition. That does not mean idealistic or charming, but one that offers enough stability to give one the presence of mind to soldier on.

These are the meek of the Earth, and you will know them by how loud they sing in a house of worship or by how much they contribute. Rather, each in his or her own way tries to make this a better world each and every day. The meek are drawn to this role because living in services to others makes one feel closer to the Creator, and the more progress they make, the more they strive for.

The bottom line here is that those entering the cusp in a state of fear will be the neediest due to the accelerant effect. They'll be lost and terrified, and they will not hesitate to frantically cling to anything or anyone they can exploit for survival. Naturally, the meek will be their preferred targets of opportunity. Here is where your role as a lifeworker will be the most painful. It is what I call survivor triage.

Survivor Triage

In wars and natural disasters, major event or battles can create overwhelming numbers of casualties. In these cases, the practice of triage serves a vital role. Triage is a process whereby limited resources are reserved for the most positive outcomes. In other words, scarcity, fate, and capability will determine who lives and who dies. Here is where lifeworkers will play a crucial role.

Many will perish in the calamities leading up to the pole shift event. Of those who survive these events, it will be the meek who find their way through the cusp with or without help. However, with the active help of lifeworkers and lightworkers, more of them will make it than would otherwise be possible. Even these extra lives represent the thinnest percentage, and that meager number could be the difference that gets humanity through a close call.

Lifeworkers, saving lives is hard enough but all things being equal, this responsibility is a heavy burden and no doubt you'd rather it never come down to that—but it may. Worse yet, you can expect to see brutal human behaviors such as you've never seen before. To illustrate the point, I wish to present a case history of a channeling session conducted in late 2008.

No One Voice Is Omnipotent

After publication of Planet X Forecast and 2012 Survival Guide, my attention turned from how we survive as individuals to how we evolve as a species. As a part of this new line of research, I followed a path first blazed by Nobel laureates who went beyond the limits of traditional science in search of answers. Like a few of them, I turned to channeling and the first thing I did was to create the following three-part channeling protocol for the effort:

1. No one voice is omnipotent.
2. Statistical trends are favored over specifics.
3. Internal logic of message is used to mitigate channel filtering.

For the channeling, we worked with several psychics from various countries around the world. In general, middle-aged women in stable and loving relationships were the most productive. The few men we interviewed were disqualified by what I call "spoiler spouses."

What most do not understand is that female psychics and clairvoyants have about a 20% chance of being in a relationship with a spouse or significant other who is supportive of their gift. The odds for a male psychic are much lower.

Consequently, spoiler spouses can be a real problem with a difficult topic such as catastrophe. The risk is that they become irrational and disruptive. Needles to say, I learned to verify the support of the spouse or significant other before conducting any interviews.

Over the 18 months of the study, psychic volunteers channeled three different types of entities: unincarnated (those who have never incarnated in human form), incarnated (those who have), and a few different extraterrestrial races.

Hundreds of hours of recorded channeling sessions confirmed that, in general, all entities contacted were benign and supportive. Also, each type showed different interest levels:

- Unincarnated entities are very good with big-picture issues and have a soft and laconic way of explaining vital ideas. They are clearly 'strategic' thinkers in every sense of the word.

- Incarnated entities are more tactical in their approach. I also found that they could empathize with certain issues due to their previous incarnations. That is very helpful.

- Extraterrestrials are more like distant cousins sending us messages of support from afar, along with poignant warnings.

With unincarnated and incarnated entities, I used a simple vetting procedure. The entity was asked to give a long-range prediction of no less than six months into the future, such as a significant natural storm or catastrophe. They always went with the weather, and I learned that predicting human behavior is just as difficult for them as it is for us.

The full study with the entity was then conducted in a trusting and impartial manner, with the final conclusions withheld until the vetting prediction was fulfilled or not. What surprised us was that all of the weather predictions were fulfilled exactly as predicted. Typhoons would show up months later exactly where they said that they would, as would hurricanes in the Atlantic.

It is also important to note that a common theme across all types and channels was a consistent warning about the possibility of our enslavement by another species.

One notable extraterrestrial entity told us that they were once, as we are now, a species divided. Because of that, they were easily enslaved by another species and many generations passed before they regained their freedom. Now, they urge young species like ours to stand together, in case the same happens to us.

All of the contact interviews, regardless of psychic or entity, were always positive and caring. The process of opening the channel is much the same with all psychics. They begin by stilling (clearing) their minds and then they put out a request, not much differently than dispatchers when they radio everyone to see who's available to take a call. Someone inevitably keys in and takes it. Unless we were working with a specific entity, that's essentially how connections with new entities were always made, through serendipity.

Once the channel with the entity is opened, the role of the psychic is akin to that of a long-distance operator. As the questioner, I addressed the entity through the psychic and the entity responded in kind.

The channeler is in a very passive role at this point, only relating questions and answers. Herein is where the issue of channel filtering must be addressed. While the psychic is indirectly involved in the dialogue, he or she can filter the message from the entity out of fear of being rejected or rebuffed.

Ask any psychic and he or she will tell you that entities, or guides as lightworkers call them, are always frank. They never couch their answers and, in some cases, questioners can react angrily. Ergo, psychics often filter the answers by saying the answer in another way, or by simply leaving it unsaid. They may also find the answer to be personally stressful and so choose not to relay it. Therefore, filtering is a concern.

However, filtering can be mitigated. Using cross-referenced questions during the first few sessions with each psychic was enough to establish a baseline. Additionally, I employed three personal interview rules that always seemed to help my volunteers feel safe in the experience:

- Keep the questions simple and avoid complex grammar and arcane words.
- Do not dwell on a topic for too long. Keep the dialog moving.
- Speak in a soft friendly voice and show appreciation at all times.

An exciting find was made near the midway point in the study that turned out to be quite useful. It happened when I asked an unincarnated entity a very direct question. It was a question a lightworker might find embarrassing, but as entities are not shy, I asked anyway: "We all appreciate your time and effort in helping us, but what's in it for you?"

Truth be known, the entities we interviewed always responded well to direct and useful questions like this. However, this answer came right out of left field for me. The answer was: "Your love makes us feel closer to the Creator."

Wow, that was useful, so after that, I added something new to the routine. After opening the channel and making the initial introductions, I would ask the other researchers monitoring the session to join me in projecting feelings of love toward the entity. The results were quite remarkable.

Our psychics would immediately report the appearance of other entities, surrounding them like moths about a light. After that, any residual hesitancy for the psychic was quelled and added rapport with the entity enabled me to cover more ground, given the same duration times.

One aspect of the sessions that was very hard on psychics involved the visualizations the entities would show them. They can often see and hear these visualizations with crisp colors and full fidelity sound.

Consequently, descriptions of predicted weather storms were fairly easy for the psychics, as long as they knew a few common TV weather terms. In a few instances, they had other sensory experiences as well.

These visualizations can be very pleasant and profound experiences in one sense, but in terms of specific scenes of catastrophe, they can be brutal. When that happens, a burnout point becomes inevitable.

The Burnout Point

The women we selected all volunteered their time and services to the study, and each came with a great sense of enthusiasm for the mission. However, it occasionally exposed them to troubling visualizations. To help you understand why, I will share one interview that is especially relevant to lifeworkers.

The visualization was so profoundly disturbing that it pushed one of our best channelers close to burnout, far sooner than expected. She's a lovely lady from Canada and, as I have a policy of protecting the identities of study participants, I will simply call her Betty.

It was the second-to-last session with Betty, one of several over a period of approximately three months. This is when the most difficult visualization documented during the study was experienced by Betty, during a channel with an incarnated entity we'd had several conversations with before, and his long-range weather forecast had been vetted.

At this time, I was researching the concept of the cusp and wanted to better understand the psychotropic effects of a period I call compression—those few months or weeks just prior to the pole shift event, where stress in the general populace builds to unimaginably dangerous levels.

What the Entity Showed Betty

The visualization started with Betty standing in the middle of a small city street somewhere near the Eastern seaboard of the United States. Or so it appeared to her. The approximate time was around dusk and she saw complete deforestation and defoliation all around what appeared to be a small city. The skies were dark and dirty with ash and grit.

Lining both sides of the street were the partial remnants of buildings and houses, and everything was covered by a thin layer of volcanic ash. We asked Betty how many people the entity was showing her and she reported that many of the adults were dead and that children were hiding in small clutches, in fear of their lives.

Then she reported seeing a man in his late 50s, thin, pale, and of medium height. His head was covered with a cloth but she could see that he had a Mediterranean complexion and that he wore two long-sleeved shirts, one over the other. He approached her from the distance, walking on the side of the street to her right, cautiously scouting for food and water.

Then a second man came around from behind her and to her left. Initially he walked along the left side of the street but then suddenly crossed the street, walking toward the older man. She could see that he was a well-built white male in his mid-20s. She was also struck by his slow but determined gait, lifeless stare, and emotionless doll's eyes.

Once the two were within arm's length, the younger man turned and violently pounced on the older man. The victim raised his arms to shield his face and the attacker grabbed hold of his right arm, swung him about, and bit into the victim's upper right arm.

With a ripping and tearing motion, the attacker then ripped a sizable chuck of muscle out of the victim's arm, sending him into shock. Thus immobilized, the attacker easily threw the victim down to the ground and pinned him. There the victim's pleas fell away. That part broke poor Betty's heart.

The attacker yanked the victim's head to one side and bit deeply into his neck, exposing the jugular vein. Then with the familiar ripping and tearing motion, he tore open the side of the victim's neck.

The entity then drew Betty closer to the victim and she saw his eyes as the life ran out of him. She did not see anger or shock, but rather a tragic sense of sadness and resignation. Then he expired and the attacker rose up and walked away without taking anything or showing any emotion at all.

Betty was stressed at this point, so I gently closed the channel. Nonetheless, she immediately insisted that she was ready to continue the work and troopered on for one more session. After that, I released her from the study and we've stayed in touch since then. With that, here is my analysis of the vision given to Betty by the entity.

The complete deforestation and defoliation she reported is consistent with a late phase 2 time frame, sometime after Earth's second transit of the Planet X dust tail and before the onset of the actual pole shift event. The thin layer of volcanic ash could be the result of ash streams from volcano eruptions in the Western half of the country.

The fact that the victim was wearing one long-sleeved shirt over another indicates that ash and soot have become a chronic problem at this point, causing skin sores and such.

The manner in which the attacker was able to approach the victim with relative ease suggests they already knew each other and that the victim may not have expected the attack. Therefore, the attacker had just recently become psychotic.

All in all, the entity's answer to my question was very direct. The visualization given to Betty showed me what I had suspected for some time—that there will be a period of compression in the months, weeks, and days leading up to the pole shift event. This will be the final stress for those who've already survived previous cataclysms.

This is the case with the attacker the entity showed Betty in the visualization. He was a young man who at one time had been a normal member of the community and likely had a friendly relationship or pleasant acquaintance with the victim. But under the stress of a prolonged series of natural disasters, he reached a breaking point where he lost his humanity altogether and morphed into a psychotic killer.

Keep in mind that in *The Kolbrin Bible,* the ancient Egyptians said the experience of the last flyby of the Destroyer (Planet X) caused men to become impotent and women to become infertile. In essence, it created a level of fear so high that it stopped human procreation.

For you as a lifeworker, the period that will be the worst will be during this stressful time of compression when people begin to go crazy with fear. Their derelict minds will concoct a bitter protest to God. In essence, "If you will not take me, then I will take those you have chosen." It will be a time of great danger.

Lifeworkers, here is where you have to shoot first and just skip the questions altogether. Once people go psychotic, like rabid dogs, they are a danger to everyone. Once the pole shift event begins, psychotics will not survive The Great Winnowing.

Those entering the cusp in a fear-based state will begin to experience a psychotropic accelerant effect of the pole shift. They will become hysterical with fear and desperate to grab hold of anything or anyone offering the remotest possibility of help.

As the accelerant effect grows, they will become completely irrational as their own fears multiply within them. When you see this, they've too far gone. The Great Winnowing will take them and they will just flicker out. This is a consistent message we received from numerous entities via several channelers.

In this time, lifeworker will be heavily affected because they operate at the PoE. Here is where moving with the natural flow of consciousness will help you to better manage the physical effects of the pole shift, like a pilot's G-suit.

So stay focused on your noble mission, keep the faith, and you'll pull through it all. Begin by identifying those capable of surviving, namely the meek, and do whatever you can to help them prevail. In return, their love and appreciation help you stay on mission.

As to the others: You'll imagine Noah sitting in his ark, as it slowly reverberates from the pounding on the hull. You'll also imagine what was going through his mind when the wailing, pounding, pleading, and screams finally began to dwindle off and then fall silent. He did what he had to do. Take heart from that.

To help you get through this, I've created the following lifeworker touchstones:

> **Each evening before you bed down**, look up toward the sky and say, "Today is finished and I am alive. It has been a good day. Tomorrow is tomorrow and it begins when I awake." Always be in the present. Past longings and future fears are expensive and deadly in the cusp.

When someone dies in your hands, tell them in a kindly tone: "Go to them. Go to the light. I am listening." Here is the message in more depth:

- Go to them: Your loved ones will be there to greet you on the other side.

- Go to the light: You'll know it when you see it, so go for it.

- I am listening: Offer yourself as a shared death experiencer.

As they pass, be prepared for a possible shared death experience if the person passing wishes to share that experience and provide you with knowledge that could help save your life and that of those who need you. Be open to it and unafraid. If possible, ask another lightworker to help you. Never forget. They keep the faith, too.

Accept the fact that your training never prepared you for a pole shift. This is when your instincts will be as important as your training. As the compression prior to the pole shift starts to happen, always remember this. When your heart disagrees with your training, follow your heart and act decisively. Do not look back for in the cusp, regrets can kill.

Each time you part company with someone you care about or respect, always say "catch you on the backside" and with a smile if possible. The backside is the aftermath following the cusp and yes, a genuine smile does add emphasis. A visual exclamation point that says you must never give up hope. Never, never, never!

Before making an important decision, tell yourself, "I must keep the faith," then count to 10 if time permits. All decisions are a gamble and the meek understand that every bit as much as lifeworkers. Squander their faith with foolish or self-serving decisions and it will cost you dearly, because they are vital to your own survival as well.

Each morning before starting the day, take a moment to ask yourself this one question: Am I a good person? Some mornings, you may not have an answer, but that is beside the point. This is because the first morning you fail to ask the question will be the day you begin to die.

If these touchstones work for you, please use them, or create touchstones of your own with the understanding that they need to help you get through The Great Winnowing. Directly after that comes the later part of the crossing called the Enlightenment.

This is when those you have protected with your own life will help you navigate your transformational experience inside the cusp in a loving way.

If you survive all this, you will be far more ready than you are today and you will be grateful, for this shall be the transformational magic of the Enlightenment.

8

The Enlightenment

The enlightened evolution of our species is approaching now, much like an oncoming passenger train. The question is: What do you want to do about it?

With this in mind, let's assume you are at the station and you see this train slowing to a stop next to the platform where you are standing. As the engine passes you, you see name painted in big golden letters. It's the Natural Flow of Consciousness Flyer, but most folks just call her The Flyer for short. The Flyer is our train to board, or to derail for that matter.

You look to see that a dust-covered ticket counter has been closed for ages. Yet, you feel a tingle in the palm of your hand and, in an instant, you're holding a bright golden boarding pass. It is a general boarding pass, which means you quickly need to choose the carriage you intend to get on. Oh, but which one? The Flyer is an immensely long train with many carriages, and each has its own conductor, urging everyone to come aboard.

Choosing a carriage is hard, but taking that first step into the carriage is harder—a big effort for some and just another step up for others. Nonetheless, when you're on the train, you're on the train.

From that point on, if you're curious about the other carriages, you're more than welcome to exercise that curiosity with a good stretch of the legs. But for the moment, you are in Marshall's carriage. Welcome aboard.

After The Flyer has left the station, you settle into a comfortable spot and look idly out the window as the scenery passes by, and the first questions that will likely come to mind pass your lips: What can we expect to find inside the cusp beyond The Great Winnowing? Will it be so beyond our current comprehension that we dare not imagine it?

This is a very good question given one weighty consideration—the degree to which we are willing to let chance decide the outcome. For this reason, my aim is to reverse the nature of this proposition until it favors good preparation over chance. Or as Louis Pasteur so aptly put it, "Chance favors the prepared mind."

So, who has got the best natural odds?

Why the Meek Have the Best Odds

The meek as originally defined have the best natural ability to survive the coming tribulation for the following three reasons:

1. The meek instinctively follow the Natural Flow of Consciousness, the pathway through the cusp.

2. There is the statistical push. When you factor out the pretenders, the truly meek become a small minority. However, the current global population gives any minority a considerable number.

3. What they experience in the cusp will seem familiar to them, but at heightened levels.

If "meek" is not you, are there any guarantees? The government whistler-blowers will tell you no and that you need to get your spiritual house in order, well in advance of the tribulation. They'll also tell you that there is no buying your way out of this and that last-minute deals and religious conversions will not work either.

In other words, only those speaking out of fear will expect a guarantee and the only one they'll receive is this: Fear is the knife's edge of death. Follow your fears and you'll inevitably stumble upon a blade.

So, cast aside imaginings and fears and begin by attempting to find a single point of truth that makes sense to you, while you still have time. Perhaps, start with something familiar.

Earlier, two different descriptions for the term "accelerant" were given. One is at the level of matter, where an accelerant is a catalyst that speeds a chemical process. Conversely, on a spiritual level, an accelerant can bring us to a stable condition of homeostasis. So, in this latter case, what does the accelerant act upon? The supernormal.

Déjà Vu

The most widely reported supernormal human experience is déjà vu. It means "already seen" in French, and it describes a very common experience one where we find ourselves in a new situation with the sense that we've been here before. These experiences are usually brief, but they can last for several minutes if the experiencer allows it.

Most have a fearful response to déjà vu, and stop the experience as quickly as possible. Afterward, they tend to avoid serious contemplation of what they've just experienced. Likewise, they will tend to hesitate to openly discuss their experiences with others.

It makes one wonder how many people are actually having these experiences and discussing them and how many are suppressed into silence.

A classic example is rape. Estimates place the percentage of unreported rapes as high as 84% where more conservative estimates place it at 60%; however, both figures are dismally atrocious. What does it say about our society when these victims do not come forward, fearing possible humiliation, shame, a mockery? A fate just as onerous as the crime itself.

In the case of common supernormal experiences such as déjà vu, the same societal mechanism of suppression comes into play, but on a much greater scale. It is a sad reality because those who do report their experiences generally express the notion that they are gentle warnings to be in a state of awareness. Some also regret not pursuing the experience so as to learn more.

Dreams

Dreamers must deal with suppression just as déjà vu experiencers, but at a significantly higher level. This is what I've personally witnessed throughout a decade of research. Time and again, people from all walks of life and from every corner of the world tell me of their dreams.

Most see the same terrible dark red skies and devastation, and many reported seeing two suns in the sky. From there, it is a witch's brew of earthquakes, tsunamis, impact events, and volcanic eruptions. I know how troubling these visions are for people, because I have experienced them too. To those of you who are having these dreams, know this:

- You're not alone.
- You can stop it.
- Let it play out.

The last point might leave you shaking your head. Why continue the experience of something troubling if you can stop it? What's to gain?

What's to Gain

It is odd how one day some can easily accept the Genesis story about how Joseph went from the dungeons to a position of great power after interpreting the pharaoh's dream about seven fat cows and seven lean cows. Then on the next day, they ridicule their spouses for having prophetic dreams of their own, dismissing the experience as nonsense.

The term that explains this dichotomy is "cognitive dissonance" a technically precise way of saying, "Don't' confuse me with facts, I've already made up my mind."

I recall one real-life account documented by a respected author about a couple on vacation. The night before a planned motorbike excursion, the husband experienced a dreadful, prophetic dream about his wife. He saw his wife in a terrible accident while riding a motorbike.

He warned his wife the next morning. Being a "rational" thinker, she dismissed her husband's dream and went ahead with her plans. Just after getting started on the ride, her motorbike malfunctioned. She crashed, flying head first over the handlebars. The resulting impact left her paralyzed from the neck down for the remainder of her life.

What struck me as odd about the account was that the author made a point of mentioning the husband's haunting regret that he had not been forceful enough to save his wife. However, what the author left out was how the wife felt about the husband's dream, now that she was a quadriplegic.

I wondered. Was she still invested in her own judgment that dismissing her husband's dream was rational and sound at the time? Would she continue to defend it despite her paralysis?

Does that sound odd? Wouldn't it be safe to assume that this question is moot as the wife must have automatically regretted her decision? No, because one must never underestimate the power of cognitive dissonance. We allow ourselves to believe that which we are prepared to believe.

This brings us full circle to the original question: What can we expect to find inside the cusp, beyond The Great Winnowing?

The answer is: What do you expect or, more to the point, what will you allow?

No matter how considered your answer may be, in the final analysis it will be a statement of how you define your own limitations. The greater those limitations are, the less prepared you will be for crossing the cusp, and here is your answer. The question itself is self-defeating and negative.

The question, "What can we expect to find inside the cusp, beyond The Great Winnowing?," is poisoned by the two little words "expect to."

Remove those words and ask the new question: What can we find inside the cusp beyond The Great Winnowing? Now, this is a question with a useful, life-saving quality.

What Can We Find?

What we'll find inside the cusp will be partly familiar and partly unfamiliar. The more familiar experiences we enter with, the easier it will be to adapt to the new and unfamiliar ones.

Delving into the greater mysteries of life is an ennobling pursuit. In a distant future, a day will come when we will see blue skies again, hear the laughter of children, and contemplate the meaning of it all.

However, what is needed today is a more practical form of spirituality. For this reason, the overarching goal is to use spiritual experiences as survival tools. On that note, let's begin with definitions.

Intellectual vs. Experiencial Definitions

A chronic problem for those first exploring their experiences is how society defines the terms. As with the word "meek," the definitions are written by people with an intellectual viewpoint, which is not to say they have ever experienced the phenomena they're defining for the benefit of those who haven't.

Hence, turning to a dictionary or any similar reference work for understanding is not necessarily going to help you connect the dots. This is not to say lightworkers are always easy to follow, even though they define terms at an experiencial level.

In terms of using spirituality as a crossing-the-cusp survival tool, four principal types of experiences are defined herein, as follows:

- Out-of-Phase: Your consciousness (spirit) slips into an out-of-phase state where time slows significantly, distractions melt away, and your problem-solving abilities are enhanced.

- Out-of-Body: Your consciousness (spirit) travels outside your body and you are ever mindful of a tenuous connection between you and your body.

- Near-Death: An event imitates the end-of-life process. Your consciousness (spirit) disconnects with the body prematurely and you are returned to the physical existence, often unwillingly.

- Shared-Death: This typically occurs when someone close to you passes and you have what could be called a second-hand near-death experience. You experience part of what your loved one is experiencing and you can communicate easily with them during and for a short time after the passing.

In the four types noted above, most experiences are positive. In a limited number of cases, those who are fear-driven can have a troubling time, especially when the experience is completely unexpected. As for myself, I initiated my first out-of-body experience when I was in college.

My First Out-of-Body Experience

At the time, I was taking a photography class and a big exam was approaching—how to use camera lens filters for black-and-white and color photography—the assigned reading was useless. The books babbled on with artistic rules, hopelessly convoluted by nearly as many exceptions. To me, it was as clear as mud and when I turned to fellow students for help, I learned they were doing no better.

A few days later, I returned home late in the afternoon from a long day on campus and lay down on the living room couch for a quick rest. It was a warm spring day in Phoenix and the roller shade was fully drawn. The shade had a tiny pinhole in the center and through it a beam of light not much wider than a pencil lead crossed over my head and painted itself on the floor.

It was quite stunning and I lay there looking at the light stream and wondered what it would be like to fly through it. If I could do that, maybe I'd learn something to help me understand how and when to use any camera lens filter.

That was running through my mind when suddenly, there I was a photon flying within that beam of light. Arrayed before me were all of the colors in this beam of light, one upon the other, like the gaily colored layers of a tall, parfait desert. I could even see spectrums that are invisible to us.

Holding out my arms, I began to fly upward through the light beam, toward the pinhole in the shade. It was easy to swoop back and forth and up and down, and fun. Swooping through the various layers of light I was engrossed in the experience and then I noticed something remarkable.

Each layer of light had a different temperature and I could feel it while swooping through the layers. That is when I had my epiphany and could finally understand why those obtuse artistic rules and their longs lists of exceptions were needlessly complex.

The answer is simple. Exposing photographic film is like a cooking process. Filters are like the knobs on our stove tops. They set the cooking temperature. Consequently, if you want to change the cooking temperature to favor a specific result with a certain film, use a filter designed to create a specific cooking temperature when you expose the film.

It was so simple that I got up straight away and went to my favorite camera store. In minutes, I bought nearly a dozen new Tiffen filters knowing precisely what each would do and with which type of film I anticipated using them. It had all clicked and I was ready for the college exam. It was an easy "A."

Know Your Limits

This first success encouraged me to experiment, and here is where my lack of preparation took me a bridge too far. If you're not properly prepared, as I learned with my third experience, you can have an unsettling experience. Consequently, after that third experiment, it was obvious that I needed a teacher to help me with attain my goals, and we met years later.

However, what I learned from the first three experiments and which has been consistent throughout my life is that when you have a genuine out-of-body (OBE) experience:

- You become aware of the consciousness connection between your spirit and your physical body via what could be called a spiritual umbilical cord.

- The memory of the experience is both enduring and vivid. Three decades later, I can still remember these first three experimental out-of-body experiences in detail.

- You open a new gateway to cosmic knowledge. Once unlocked, this door can swing both ways.

- While in a state of love, you are in control of the experience and can terminate it at will.

- Preparation greatly enhances your prospect of a positive and ultimately useful experience.

For those lifeworkers who wonder if they can have an out-of-body experience, let me add one final note. At the time of my first OBE experiment (flying through the light beam), I was serving in the Army National Guard. I had a great job. I was a photojournalist for the commanding general.

Illustration 53: Marshall Masters (1976)

Between the first and second experiment, I was assigned to photograph the wreckage from a 1956 Grand Canyon mid-air collision. My job was to provide photos of the larger parts of the wreckage for special handling. We choppered in and I saw the Grand Canyon with my feet dangling off the side of a Huey, flying through just feet off the water. Then we arrived at the crash sites.

To this day, I can still feel the emotion that had been fused into the wreckage of those two doomed airlines. It was a result of the first experience of flying through the light beam. I became sensitized and remain so to this day. To put it in perspective, I'll share what I felt at the crash site, but first, a little background.

It was nighttime over the Grand Canyon when a United Airlines DC-7 began climbing through the clouds to smoother air higher up. The pilots suddenly saw a TWA Super Constellation dead ahead. They turned to avoid a collision but it was too late. Their left wing tip sheared off the empennage (the rear of the airplane) rendering it completely uncontrollable.

After the collision, the Super Constellation immediately flipped over on its back and dropped in a flat spiral. It hit the top of a sloping hill and tore itself to pieces for about a half a mile until the wreckage came to a stop near the edge of the canyon.

As I walked through the wreckage, I could see the cabin inside the airplane was a chaos of wind and darkness, people screaming hysterically knowing their doom was certain. I could hear it, especially the women. All the fun of flying through the canyon just off the deck faded. This was sobering.

From there we went to the second crash site. The DC-7 had augured into a hard rock face and all that was left was a large molten pool of cooled aluminum. It looked like a huge dentist had given the hill a large aluminum filling. As I walked through that wreckage I sensed something very different. I could see in my mind hands clenching the armrests and I heard quiet prayers and crying. I knew these people died still thinking they had a slight chance.

I shared this with the Huey pilots and one said, "That makes sense. We found one of the props and, by the pitch of the blades, we could see that they were fighting to make it." That got the pilots to thinking and before we left they told me they figured that if the DC-7 had passed a hundred feet to the right of the rock face, the pilots might have been able to make a controlled crash landing. I was glad to leave that place.

If you've never experienced an OBE, what are the chances you will? That depends on you, but one thing is for certain. You'll have a 100% chance of one of two things happening to you when you cross the cusp: You will have an OBE or you will likely die.

In the meantime, the most likely first spiritual experience a lifeworker will have is what I call an out-of-phase experience.

Out-of-Phase Experience

Of the various types of spiritual experiences, the out-of-phase type is the most cerebral. Given that lifeworkers will be zoned in a point of equilibrium, this type of experience will be a very useful tool for anyone in survival mode.

This is because you are not going outside your body either with or without a tethered connection. Rather, you stay within what could be considered the first moment of a near-death experience. In this state, your consciousness is out of phase with the natural world around you. Time shifts into slow gear. A lyric from the Wreck of the Edmund Fitzgerald, by

Gordon Lightfoot, says it well, "Does anyone know where the love of God goes and when the waves turn the minutes to hours?"

I know personally this to be true because one such experience saved my life another experience from my college days. It happened while returning from Prescott, Arizona, with another Honda 350 motorcycle rider. His bike was designed for off-road riding and mine was a street variant of the same model.

Illustration 54: Marshall Masters (1975)

We were weaving through the hill country toward the valley area on a narrow two-lane road with few guard rails. My riding friend got well ahead of me and I raced ahead to catch up and felt my speed was low enough for an approaching hairpin curve. There was no guard rail, just three inches of sand and gravel beyond the asphalt and a zippy 800-foot drop after that.

The curve was really tight and I banked hard to clear it. That's when my bright idea to install a roll bar on my motorcycle put me in a real jam as it starting dragging along the road, lifting the front wheel off the ground. At that point, I was out of control and barreling in a straight line toward an 800-foot fall some 50 feet away.

The instant I realized that, my only thought was that I needed to figure this out, and fast. Then suddenly, I was out of phase and time moved so slowly it seemed as though it was stuck somewhere between tick and toc. Immediately, I set my mind in motion to run down the options and it went on automatic, performing calculations entirely on its own.

As each calculation finished, I reviewed the result. The first was quick. If I continued as is, both I and the motorcycle would soon be airborne. Non-starter.

The next option was kicking the motorcycle out from underneath me. That came back negative as well. My momentum would continue to carry me forward and I would still become airborne, not far beyond the motorcycle. So much for the easy ones. Now I had to dig.

Then I remembered two weeks earlier, when I installed a new set of tires with a soft tread on my motorcycle. Designed for optimal traction they lasted half as long as other tires, but I wanted the traction. So now, could the softer tread of these newer tires make a difference?

I plugged that into my brain calculator and back came a plan in an instant: Regain control of the motorcycle so that I could steer my way through the curve. With a plan in hand, I made the choice to act.

The instant I did that, I dropped back into real time and in a single motion, removed my right hand from the throttle, cocking it backward. Then I slammed the palm of my hand forward into the throttle and handlebar as hard as I could.

Once I'd removed my hand from the throttle, the engine began losing power, and by slamming a handlebar, I'd jolted the front tire onto the roadway. Here is where that slightly softer tire emulsion paid huge dividends. The tire bit the road and pulled me around the curve to a straightaway, where I slowed to a stop alongside the road.

When my friend pulled up, I was trying to light a cigarette but was shaking too hard to light it. He lit my cigarette for me and said that he had watched everything from a vantage point above us. He said he was certain that he was going to watch me die that day and then asked me how close I thought I had come to the edge.

I shrugged, and he said he saw my rear tire kicking up a rooster tail of sand and gravel from the three-inch berm at the edge of the road. It was that close.

What I learned from that experience is that:

- An out-of-phase experience is a very powerful, problem-solving cerebral accelerant.
- You must be actively seeking a solution to initiate an out-of-phase experience.
- Follow your instincts and when you have a workable solution, take action.
- Trust your own calculations, even though you do not know how they are derived.
- You cannot have a useful out-of-phase experience while in a state of fear.

What is really important to remember is that your initial response to a threatening situation makes all the difference, and fear is the killer. Ergo, if you want an out-of-phase experience to work for you in a life or death situation, you must stay so that it can it work for you. Most importantly, you must ask for it.

It is also possible for an out-of-phase experience to overlap with what are now described as shared-death experiences.

Shared-Death Experiences

Much has been written about these topics, but in the last decade or so, we've seen a new kind of research from surgeons and physicians specializing in grief counseling and hospice care. They are reporting something that is happening more frequently now to their patients and their loved ones, concurrent with the rising level of consciousness within the greater public. They're called shared-death experiences, where those close to the person dying share part of the death experience.

This is a unique type of out-of-body experience, and I personally experienced one the day my beloved father-in-law Aron passed away. At the time, my wife's parents were visiting us.

Aron was a kind soul, with a weakened body. Years of diabetes and heart problems had taken their toll, and just after six o'clock in the morning, I was awoken by the horrific screams of my wife and her mother.

I threw on a robe and ran upstairs and quickly sized up the situation. To my left, two hysterical women running back and forth screaming. To my right, Aron's lifeless body on the floor of my upstairs bathroom. He had given himself an injection of insulin and had a massive coronary.

I told my wife to call 911 and quickly went to check his vitals. Nothing. Absolutely nothing. He was gone. That's when my training kicked in.

Before becoming a photojournalist in the Army, I had been a medic, rated 91-B, which was roughly equivalent to a licensed practical nurse at that time. It never fails to impress me how, after all those years, the training could still kick in. Something to be grateful for.

First, I cleared his airway and then gave him a hard rap on the sternum. Nothing. I rapped him again. Still nothing. From there, it would be compressions until the paramedics arrived and that was iffy at best. Then I gave him a third rap and with that, he drew a very small, raspy breath. Something! I checked for a pulse. It was thready, but at least something was there.

By this time, my wife handed me the phone and while I was talking to the 911 operator, his breathing stopped again. I tossed the phone back to my wife who resumed screaming at the operator and shut the bathroom door, though it did little to muffle the sound.

Then I stopped to collect my thoughts. At this point I was just frustrated and shouted at him, "Dammit Aron, work with me!" The next thing I knew, I was out of phase and time slowed, but not to the extent it had during my motorcycle experience decades earlier in the hills of Prescott, Arizona.

For me, it was instant relief as the hysterical screaming was not a distraction. I could faintly hear it behind me, like a distant buzz, but that was all. Now I could hear myself think and Aron was with me, talking to me, even though the man on the floor before me was unconscious and holding on by the barest thread of life. Yet, we were together and we began working on the problem.

By profession, Aron was an aeronautical engineer and at the time I was working in the computer industry, so there really wasn't much of anything spiritual about it. Just two geeks trying to think out a way to get him stabilized for the paramedics. The interesting thing was that Aron didn't speak a word of English and my corresponding language skills were pitiful. Yet we spoke, or more accurately, thought together with perfect clarity of meaning and intent.

While most people report not wishing to return to their physical bodies during a near-death experience, Aron was fighting and clawing his way back for everything he was worth. Not because his physical existence was enjoyable. Years of diabetes and heart troubles had ruined all that. His physical life was filled with pain, but for the joy of being with his girls, it didn't matter to him. Aron was coming back for his girls.

For him, his wife and daughter were the most precious things in his universe, and he was not about to casually let go of them. Any time and moment of embrace; one more smile; one more gentle kiss that was all he wanted now. It was so strong within him that longing to embrace.

His pulse was still thready so I told him the paramedics would arrive very soon, and we decided to work on restarting his breathing until they arrived. That was when he asked of me an unusual request that I show his wife how to give him mouth-to-mouth resuscitation.

I said I could do it more effectively myself but he insisted. "Show her," and with that I was back in phase, replete with my wife's hysterical screaming. However, my mother-in-law was more focused now. Gesturing her to kneel by his side, I began showing her the procedure.

The first time she placed her mouth on his and blew, it was amazing. To my utter surprise, he took a long, deep, raspy breath and his breathing resumed. Things were still dicey as his pulse remained thready, but a little bit of color was returning to his face. At that point, I could hear the ambulance pulling up to our front door.

The paramedics worked on him until they felt he was stable enough for transport and took him to the local hospital. As they left, they said he had a good chance of making it. Things began to look a little hopeful. Perhaps the plan was working. Perhaps.

What I'd experienced with Aron was life affirming, but the hysterical fear of my wife and mother-in-law were life draining, and I was a bit punch-drunk from it. A quick shower restored me and we set off for the hospital.

My wife and mother-in-law sat in the back, leaving the front passenger seat empty Aron's usual spot. It was an eerily quiet trip and finally we curved around the exit ramp and turned toward the hospital. As we were crossing back over the freeway, a brilliant golden ball appeared, floating over the front passenger seat. I immediately felt Aron's glowing presence and knew that he'd passed on.

The car was filled with golden light and a tremendous outpouring of love. His radiance was so intense that I had to slow down and hug the side, concerned that I'd cause an accident, and then he faded away.

I looked back at my wife and mother-in-law and asked, "Did you see that?" They looked at me like I was talking nonsense.

At that point, I knew how things were about to play out and those last few blocks to the emergency room were probably the longest moments of my life. Of course, the nurses steered us to small waiting room instead of escorting us bedside.

It was all beginning to sink in when the doctor entered the room. He explained that they'd tried their best, but that they couldn't restore a normal heartbeat and lost him. In the end, Aron's rusty old pump got him, but at least we succeeded in giving him his last wish—to feel his wife's lips upon his just one last time. Something like that teaches you a lot about love.

A few weeks later, in mid-August, I was walking through the local old-growth redwood grove near our home. We had visited that park with Aron the day before he died and I remembered taking a picture of him leaning up against a very old tree.

As I came upon that same tree, I stopped and placed my hand on the spot he'd leaned against. Then once again, I was out of phase and there he was, appearing as a younger, more vital man in his prime.

Joining him were my wife's relatives from the other side, and they asked me to give my wife and mother-in-law a simple message, "We'll all be here waiting for you when you come." They kept repeating themselves until I agreed to relay the message.

Then they gathered close around me and so much energy passed through me that for a brief moment I thought I would begin to levitate—and I'm no tiny fellow. It was an amazing experience, which I later learned happens quite frequently.

During that incredible experience, Aron also gave me a prediction. He said, "The world will soon begin a terrible tribulation." Those were his exact words and a few weeks later, the World Trade Center twin towers in Lower Manhattan collapsed at the speed of gravity, straight down into their own basements. It was 9-11 and we've been in the tribulation of war ever since.

The Survival Benefits of a Shared-Death Experience

What I learned is that shared-death experiences can offer those with open minds and stout hearts a unique survival advantage. When a loved one passes, he or she will become privy to a much broader view of things to come and, given a chance, will share with you.

To help visualize this, imagine you're helping to launch a manned observation balloon. As the balloon rises upward, the observer begins to see far distances and shouts something short and simple to you like, "Danger is to the East. Food and water are to the Southwest."

As the balloon continues to rise, the sound of his calls grows faint and his other messages are unclear. Nonetheless, you now have options you didn't have before. Ergo, death is what we choose to make of it. It can be a selfish time of denial and rage or we can share it with love and humility via a shared-death experience.

During the coming tribulation, the sheer volume of shared-death experiences will be more powerful than we can imagine. These experiences will play a critical role in the survival of small clans of people as well as a transformative role in terms of social evolution.

No doubt some will also call it untested, but this phenomenon has already been witnessed and heavily documented, by something I like to call the "Oh My Johnny" scenes.

Oh My Johnny

During WWII wives and mothers all across the nation were having shared-death experiences with their sons and husbands, and it was very troubling for the women and bad for morale. Hollywood's answer was to include an "Oh My Johnny" scene in several films of the day.

This scene was a typical Hollywood formula. Given that these men were dying on far-flung battlefields, women were typically having these experiences in the middle of the night. Consequently, the scene always opened in a dark bedroom on a moonlit night.

In the bed, a woman is sleeping when all of a sudden, she wakes and lurches upward screaming something like, "Oh my Johnny. My Johnny is dead. They've killed my Johnny." The scene would close with the woman sobbing hysterically.

The whole point of the "Oh My Johnny" scene was to entertain audiences. It was to tell women that they were having a normal experience and that there was nothing to be ashamed of. The meaning was clear. Because it was being shown in the theater, they could talk about it with others who were seeing it happen in the movies as well.

Granted, the scenes were done for wartime morale, but the real benefit was that it helped these women, in some measure, by removing a needless burden of shame as a result of their shared-death experiences. At this point, be honest with yourself. Are you willing to discount a phenomenon that has been documented for well over a half a century? Or, can you now see how shared-death experiences serve to increase the odds of survival?

If so, begin preparing yourself for these out-of-phase and shared-death experiences today, and there are several things you can do to hone these spirituality survival skills for the future.

Fear-based Expectations

During the tribulation, you'll pray enough for several lifetimes so the first step in preparing yourself begins with a frank examination of your current belief system.

Do you live with the expectation that those who do not share your faith will suffer terribly for it? That your beliefs will be vindicated in the sufferings of others? That you will be rewarded in some way for professing a certain belief? That you were not born with these beliefs, but rather were taught them?

A lyric from the show tune "You've Got to Be Carefully Taught" from the 1949 Rodgers and Hammerstein musical *South Pacific* says it best:

```
You've got to be taught before it's too late,
Before you are six or seven or eight,
To hate all the people your relatives hate,
You've got to be carefully taught!
```

Whether this is how you were taught or how you chose to define yourself makes no difference, your chances of crossing the cusp will be slim to none. This is because crossing the cusp is not a day of judgment, it is simply an evolutionary threshold event you must pass through—or not.

You may feel rightfully entitled to hold these beliefs born of fear as expressed by various shades of triumphalism, hatred, vengeance, and the like. This is because an expectation of judgment and punishment is fear-based.

You can pile on all the rationalizations and justifications you want in furtherance of your case, but in the end, you're fear-tainted goods. This is because any shade of fear resists and runs contrary to the Natural Flow of Consciousness. Consequently, once inside the cusp of the universe, the flow will not work against you. Rather, you'll miss the train called The Flyer, so to speak and be left at the station.

Remember back at the beginning of this chapter; the dust-covered ticket counter closed for ages?

If you are tainted with fear, you'll remember how the boarding passes suddenly materialized in the hands of some, but not others. Why?

There is a reason for that. The boarding passes for The Flyer represent a threshold event and can only materialize in the hands of someone who is already in the Natural Flow of Consciousness. In a manner of speaking, it is like applying for a position that requires a college degree as evidence of your commitment to personal goals. You must meet the entry requirement to be considered. However, crossing the cusp will not require a college degree, but it will require that level of commitment. It's just as simple as that.

Therefore, if you are waiting for someone else to suffer for your own fear-tainted beliefs, you'll be left at the station with whatever is in your suitcase to get you through the cusp. If all you have are your own fear-tainted beliefs and the element of chance, you're not going too much farther.

But you ask, fear is a part of us so doesn't it make sense that a little bit of fear is good? Isn't it true that we need fear to survive? Wouldn't it be more sensible to learn how to channel these fears? All these are cloudy questions and nothing more.

To help visualize this, imagine a crystal clear glass filled with pure spring water sitting on a sunlit table.

This is how we are born into the world, absent of fear-tainted beliefs such as hatred, vengeance, and triumphalism.

Now in your hand you hold a small glass vial filled with an ominous black mixture labeled "Fears – Various." Using a dropper, you gently place one small drop of this black ink upon the surface of the water and stand back.

The light passing through crystal glass highlights brackish fingers, snaking through the clear pure water in the glass. In the span of a few breaths, the glass becomes fully tainted and what was once a beauty to behold becomes cloudy, dark, and unpalatable. So, is that the way it has to be? No!

What you need to do is to filter the waters of your own beliefs. You know how the process works. More of us filter our own water today than ever before because we want to control what goes into our bodies. The case for spiritual filtering is exactly the same. So is the result: crystal clear water.

However, filtering your spiritual waters is a time-dependent process and quite difficult if not impossible to do when the world about you is crumbling in a cataclysm.

Once inside the cusp, your own fear-tainted beliefs and expectations will draw you backward, against the Natural Flow of Consciousness, and at a geometric rate. The closer to raw fear you are, the faster you go. Consequently, early on in The Great Winnowing, you will retrograde to the point of your own death.

The truth is as simple to read as a billboard. Look inside and be honest. None of us are saints, so be realistic enough to find your own sense of equilibrium. This does not mean that you disavow your religious beliefs, for almost all religions profess a way to go with the Natural Flow of Consciousness. Likewise, they can also offer fear-tainted interpretations that serve material aims, because beliefs systems are what we chose to make of them.

Therefore, find those elements of your faith or beliefs that go with the flow and filter out the others. And if need be, have the courage to go it alone when others mock and ridicule you as a heretic. Above all, you must be committed, even if that means standing alone in your knowing and in your integrity.

Another fear you must deal directly is the fear of your own mortality.

The Fear of Death

In Elbert County, Georgia, USA, is granite monument that many believe was put there by Freemasons due to its precise celestial alignments. Called the Georgia Guidestones, you could say it is a modern, secular version of the Ten Commandments.

The first inscription tells us: "Maintain humanity under 500,000,000 in perpetual balance with nature." The other nine follow in general support of the first and in a manner similar to another anonymous message to humanity given us by the Rosetta Stone.

Carved in 196 B.C. by Egyptian priests to honor a pharaoh, it was discovered in 1799 by French soldiers rebuilding a fort in a town named Rosetta (Rashid) in Egypt.

The Rosetta Stone was a breakthrough discovery because it gave scholars a cross-reference between the Egyptian and Greek languages, using three different scripts—hieroglyphic, demotic, and Greek. It is why this black, basalt stone, roughly the size of a small coffee table, unlocked the key to understanding ancient Egyptian hieroglyphics, such as those inscribed in the Great Pyramids.

Conversely, the Georgia Guidestones, which are also called the "American Stonehenge," is nearly 20 feet (6.1 m) tall and comprises six granite slabs, collectively weighing more than 240,000 pounds (110,000 kg). Like the ancient Rossetta Stone, the George Guidestone is also multilingual, but with 8 different translations of the same 10 inscriptions in English, Spanish, Swahili, Hindi, Hebrew, Arabic, Chinese, and Russian.

When this monument was erected in March 1980, the global population was approximately 4.5 billion, and now it is expected to cross the 7 billion mark in 2012.

Given that the Georgia Guidestones predict a post-historic global population of just 0.5 billion, as the population continues to increase, the individual odds of surviving are likewise dwindling. Do the math yourself and if you choose to better those odds for yourself, you need to tackle your own fear of death.

Today Is a Good Day to Die

As Western thinkers, we need to understand that our predominant belief systems were all created during a time of relative quiescent given the geological scale of Catastrophism. Consequently, they are dramatically different from the belief systems of ancient indigenous peoples, and the misunderstandings are fundamental.

For example, in the 1950s, western movies and TV shows became the thing in America. They were so skewed that native Americans refused to appear in them, no matter how much money was offered. Consequently, a gaggle of white guys would smear a red tint all over their bodies (that could take days to wear off) and speak in the Western tongue of Hollywoodhaha in gravel-like tones. It worked for most audiences.

Of course, there was always that moment in the film before the big charge when the Indian chief would say, "Today is a good day to die." For audiences, this suicidal gesture of kamikaze-like bravado signaled that now is not a good time to get popcorn. The good stuff is coming.

For native Americans, the movie line "today is a good day to die" is an out-of-context sound bite that glorifies kamikaze-like bravado, with aborted snack bar runs.

When a native American says "today is a good day to die," it is the last part of a broader statement that essentially says, "I have accomplished most of what I set out do to in this life. I have experienced love and I have always tried to do my best. If I must fall here and now, then today is a good day to die."

As with all such systems, no matter how they start out, a self-interested few will inevitably co-opt them to suit their own aims. They are the ones who control perception and this is a mighty power. However, consciousness can filter this out.

Look at what we have today—Westerns that are far more realistic and respectful of indigenous peoples. Consequently, when we see red-painted white men ride across the screen, it will be a parody of the 1950s Westerns, such as Mel Brook's comedy, *Blazing Saddles* (1974).

Granted, a lot of people worked hard to make this consciousness come about, but the fact remains that this massive shift in public consciousness took just a few decades. This is because this awareness followed the Natural Flow of Consciousness.

The point here is that the game of life is not rigged in favor of elites and their minions where you'll have to dance to their tune one way or another. They'll certainly want you to believe this, and what they can do with their smoke and mirrors is quite impressive.

There is only one reason why people live in fear of death. It serves the worldly needs of a few. That is all.

The Ancient Right of Monarchs

Our present concept of death in the West is a carryover of the ancient right of monarchs to own the land and all the souls upon it. Ergo, as a vassal or serf, your life can only end in one of two ways: by natural causes or when it serves the interest of the monarchy.

To enforce this right of monarchs, they concocted fear-based doctrines and, with the help of those with similar interests, indoctrinated the masses so as suppress awareness of their own natural rights. Though we may be cheated of them, however, we have the natural rights to make determinations about our life, and they are ours alone.

Obviously, this practice of suppression through indoctrination serves the interests of the elites, but given that, what do elites themselves fear? They fear the awareness of the people, because when it reaches critical mass, it's payback time and they lose.

When that control starts to slip, their use of smoke and mirrors ratchets up into headlines warning of certain moral decline and social chaos should the state return this natural right to the people. There will also be other drumbeats, such as suicides mushrooming out of control as long queues of unhappy people flood into assisted-suicide centers. A grim picture it is, but is it an accurate picture?

When Oregon passed the Death with Dignity Act in 1997, it became America's first assisted-suicide state. Have suicides skyrocketed in Oregon since then? No, they're at the same levels as those states which continue to treat assisted suicide as homicide.

Are assisted-suicide caregivers in Oregon flooded with unhappy people, stupidly seeking death as an escape from a sad or disappointing life? No, the people are those who exercise their natural right under the law and those meeting the spirit of the law—to help those with incurable, painful, and terminal medical conditions.

The result in Oregon is a compassionate and humane solution that works. In just over a decade, it has shattered the mirrors and blown away the smoke of elitist propaganda. It is because this brave political solution follows the Natural Flow of Consciousness.

The point here is that no matter what anyone else tells you, whether it be from a pulpit, courtroom, the floor of Congress, or the joker at the end of the lunch counter for that matter, the Creator made us all smart to figure this out on our own.

Do we need to go to college or listen to a sermon to get started? That's like eating two-day-old sandwiches for lunch. A lot of dry munching that's hard to swallow. There are better ways.

Taking the First Step

A good first step is to enroll in a local volunteer hospice caregiver certification course or to volunteer your services to a local hospice. Once you complete your training, you will see the end of life from a new perspective. This in turn helps you to understand the process well enough to gain confidence in your own ability to have positive, shared-death experiences to help others.

Even if you never volunteer one hour in a local hospice, you will learn about and possibly experience what really happens during the end-of-life process. Therefore, a course of volunteer hospice training offers a wonderful benefit for good students. They will complete their studies better equipped to survive the tribulation.

Likewise, expect one last elitist attempt to re-infect you with the fear of mortality. The elite will issue proclamations through their minions that shared-death experiences exploit the dying. Before you choose to embrace that proclamation, become a hospice caregiver and embrace a dying person, and ask him or her about it. The dying person's thoughts on the matter will be your true answer.

If possible, discuss the topic of death and dying with close friends and family members in an open and non-judgmental manner. As you do, be careful of a syndrome called "the doors fly open" by hospice caregivers. It describes how fear of mortality can trigger a sudden, emotional rage.

The term uses a hallway closet full of fears and regrets as the basis of the analogy. Some of these fears and regrets in the closet are of our own making, but most are the result of indoctrination.

The longer we live and the more regrets we have, the more the closet door bulges outward. Then when death reminds us of our own mortality, the latch is shattered and the closet door flies open, spilling out its fear-tainted contents into the hallway to the detriment of all.

If you attempt to open a dialog with loved ones and they show signs of irritation or discomfort, go no further. You will only provoke an outburst of "the doors fly open," and it will be a very unpleasant and counterproductive experience for all concerned. This is why you must never attempt to proselytize this awareness. Rather, move on and, in your own quiet way, strive each day to heighten your own natural abilities of observation and contemplation.

Observation and Contemplation

In the times to come, you must focus your powers of observation and contemplation as never before. It is not so much about what you discover, but that you hone these natural skills, which increasingly tend to languish in our modern world.

If you've yet to have a supernormal experience, be patient, you will. In fact, you will face many new spiritual unknowns in the cusp. If you succeed, they will be new and transformative spiritual experiences.

Therefore, do not fixate on what you will experience in the future, as you already possess the natural talents needed: observation and contemplation. These are the best tools in your kit for quickly understanding new unknowns and how to deal with them.

To make the most of them, you need to be able to observe as much as possible during the experience and then to use these stored memories for post-event contemplation. One benefit of these two natural mechanisms is how they help us filter out the garish noises of the world. In a calmer mind, one has a freer hand to seek out wisdom.

Think of fear and all its permutations as desert sands. Like a mighty storm, the coming tribulation will whip these sands up into the air, blocking the horizon of wisdom from sight. The greater your powers of observation and contemplation are, the sooner the storm will pass over you, revealing the far horizon of wisdom once again. So, how do we recognize this survival wisdom? Like Aesop's owls.

The Owls of Aesop's Fables

When we hear the terms "wisdom," "observation," and "contemplation," ancient Greek philosophers immediately come to mind, so this is as good a place as any to begin. In early Indian folklore, owls were viewed as symbols of wisdom and helpfulness, with the power of prophecy—a theme that carried forward into Greek myths and beliefs.

Through the centuries, parents have enjoyed reading stories from Aesop's fables to their children. This also happens to be an excellent way to learn why the ancient Greeks viewed owls as symbols of wisdom.

What made the ancients draw the conclusion in the first place that owls represent wisdom? The answer begins with how most people recognize wise individuals. It is the manner in which they observe those about them. For example, they use the art of listening.

In general, a wise person will listen carefully and perhaps ask confirming questions before drawing a conclusion. Conversely, those who impress others as unwise will make one or more of the following listening mistakes:

- **Anticipating:** This is the most common mistake, where listeners hear just enough to draw an incomplete conclusion about what the speaker is talking about. Then they stop listening to prepare their response. However, when questioned about what the speaker has just said, they can appear to be clueless. This is a powerful way to make someone look foolish.

- **Clipping:** This behavior usually follows anticipation. Listeners are so eager to respond, they begin doing so as speakers deliver their last few words. This is a disrespectful gesture toward the speaker that tends to polarize the discussion.

- **Interrupting:** This is anticipation with attitude. Listeners interrupt the speaker so as to shift the attention to themselves and what they want to say. Unless the point they're making is worth it, these tactics are viewed as a disrespectful gesture to everyone with an interest in the dialog.

In this example, the wise person will show respect for the speaker by listening objectively and intently to what the speaker is saying. Then after the speaker completes the statement, the wise listener will pause briefly to consider a response. Even if the answer is lacking, it will nonetheless be viewed as respectful.

Therefore, what makes the wise person in this example appear to be wise has less to do with actual wisdom and more to do with how the wise employ their own powers of observation. In this regard, evolution gave the owls a physiology that makes them one of nature's most sophisticated masters of observation. In other words, they look the part.

Consider this: The eyes of an owl are so powerful that if we humans had similar capabilities, our eyes would be the size of small oranges. Of equal importance are the feathered cups around an owl's eyes, as they serve an important purpose as well.

The eye cups capture line-of-sight sounds from whatever the owl is focused on and direct these sounds into the owl's ears.

In terms of audible ranges, owls are comparable to humans. However, their hearing is finely tuned for certain frequencies which enable them to hear slight movements, such as a mouse foraging on a forest floor. Yet, that's only the half of it. Owls' ears are also asymmetrical. One is always a little higher on one side than the other.

This asymmetrical positioning allows the owl's brain to compare the differences between its two ears, so it can instantly determine the precise range to its target. In other words, owls automatically hear what they see and can track their targets with accurate, real-time, targeting parameters.

This is why the ancients viewed the owls are symbols of wisdom; they knew that wisdom begins with observation. From observation comes contemplation and, from that, wisdom becomes possible.

In terms of crossing the cusp, your own powers of observation and contemplation will be valuable tools for spiritual survival, especially during The Great Winnowing.

Also know this. If you have not already experienced a genuine near-death or major out-of-body experience, you will experience one inside the cusp. These experiences will be unavoidable, and new ones must also be anticipated. All these will happen frequently, perhaps several times a day, and they will be profoundly transformative in ways we can only surmise at present.

Transformative Experiences

All the previously discussed experiences will certainly play a role in spurring an enlightened social evolution of our species. However, genuine, major, out-of-body and near-death experiences will be the most transformative aspect of the enlightenment on an individual level—once inside the cusp—and the most calming.

A major event is one that transforms the experiencer. For example, two people may tell you they've had an out-of-body experience. Assuming you believe they are being honest with you, determining if one or both actually had a major out-of-body experience is simple. Ask them, "Are you afraid of death?"

The first one considers the question for a moment and then responds with something like, "What an odd question. Why are you asking that?" At face value, this is a prudent, logical question.

However, the second answers, "No. Not anymore, but that's not to say I wish to die badly, which I certainly do not." This second experiencer is one who was transformed as evidenced by his or her losing all fear of death. Conversely, while the first one may have had a profound experience, it was clearly not a transformative one.

Therefore, not all out-of-body experiences are profound [transformative], but nearly all near-death experiences are.

Those who have had a transformative experience become less interested in material issues and self-serving goals. They begin to strive for balance through harmony and ways to live in service to others that give their own lives meaning.

In other words, the transformation moves you forward within the Natural Flow of Consciousness—in some cases, dramatically so.

This is the same process for all life, and many of the enlightened guides [entities] we interviewed over a period of 18 months during our channeling study made it a point to stress that fact. What we repeatedly heard was the same essential message:

They too are advancing with the flow of consciousness toward the Creator and there is no hierarchy, or pecking order, if you will. To briefly return to the previous analogy of The Flyer, they too see themselves riding on same train and they see one class of service—service to others.

From the end of The Great Winnowing through the enlightenment within the cusp, more human beings will be having simultaneous transformative experiences than ever in the entire history of our species. This could become a one-of-a-kind, globally shared experience that propels our evolution along both social and physical lines.

For this reason, virtually all those who cross the cusp to the aftermath on the backside will emerge forever changed. Free of indoctrination, none will fear death ever again, nor will they be interested in creating another unsustainable world based on greed. To this they will say, "Never again."

Nonetheless, the elites and their minions who do survive through technical means will continue as before to anchor the foundations of another unsustainable fear-based world in their own image. However, they will also be the conveyors of technology from one time to the next. In this they can serve a purpose to the good, but only if this technology is used to help us become a peaceful, stellar species.

Previously I said that this evolution favors the meek but that it is ours to lose. This is where the meek stand to loose all they will have gained, unless they are vigilant and determined not to submit. This shall be the moment when the surviving elites reassert their technology.

In this future time, the old will confront the new and how those transformed in the cusp handle that will largely determine the fate of humanity.

Here is where I believe that those who are transformed within the cusp will be given a very special Gnosis [a highly spiritual understanding] and the guides will share this knowledge when they believe the time is right, and only then. It is their way, and yes, often we feel as though we're the last to know. But know it we shall. Of this I am confident.

In the meantime, I know how difficult it can be for those who have never had a transformative experience to have such confidence in the future. The intellectual can so easily fail us.

With that in mind, I will share my own transformative experience and then you shall hear directly from the guides yourself. Please remember that for the moment you're sitting in Marshall's carriage on The Flyer, so a good stretch of the legs is all that's needed to explore different thoughts and ideas.

Nonetheless, while you are here, this sharing may help to give you the sense of a transformative experience if you have not yet had one. Feel free to use this information to best serve your needs.

9

Transformation

Lightworkers and those who have already had a transformative experience often help those who've yet to have one cross this threshold by recounting their own experiences and those they've witnessed. This helps in part to give future experiencers a sense of what they themselves may encounter. It is a part of paying forward, and I do it now and for one reason only —I'm in it for the species.

Those who would lead you to believe that we are all alone in the universe are spreading orparroting a perverse falsehood, because we are not alone. Likewise, we have never been alone and we never will be alone. Think back to the first part of this book, where anonymous friends from afar went to great effort to give us an urgent warning via the Avebury 2008 formation.

In this chapter, you will receive powerful messages from other friends from afar via 12 readings I personally channeled from enlightened guides called The Elohim. These guides and countless others like them are wonderfully kind, incredibly patient, and connected to us in a very loving way.

They are doing all they can to help us evolve, and they do not lose their resolve, question their commitment, or throw their hands up in frustration with our foolishness and just storm off. That's not who they are. Not in the least. Their style is more what you could call pleasant persistence, and they move the ball down the field every chance they get.

In a manner of speaking, this is a game like the ancients played, but with the highest stakes imaginable. This time, the prize is our survival as a new stellar species. As to second prize, there is no second prize, that is, unless you consider slavery to be a prize. Otherwise, go team go.

This is why the guides warn us time and again that while this is our time of evolution, it's also a time that's ours to lose. Can we prevail? Of course we can! Moses proved that much. He used the chaos of the last flyby to free his people. This time, it can work again, for everyone. We can do this.

With that, I share my own story as an example for those who may wonder about what they may find in the cusp.

My First Transformative Experience

After my third astral projection experiment while in college, I knew two things. First, I needed to be better prepared for my goals. That meant getting a teacher or mentor, so I put out a request to the universe and four years later, he showed up.

His name was Hogue and he was a remarkable fellow. Six foot two with the build of a man used to squeezing a living out of a hill country ranch, he had the most remarkable smile of any man I've ever met. It was like God started with his smile and then painted a man on it.

He mentored me, immersing me in the ways of the Medicine Wheel. After several months had passed, he nonchalantly asked if I had a place in mind for my vision quest. I told him I did and he simply said, "Then you're ready."

The following month I traveled to that very place, a twin saddle between two large peaked hills, deep inside the Prescott National Forest in Arizona. There, the quest began at a small cabin near the base. There, I prepared for a day and a half, as I'd been taught.

The morning of my vision quest, I arose with calm resolve. Taking nothing more than a canteen of water and a pistol loaded with snake shot, I hiked up to the saddle and found my spot, just where I imagined it would be, in the middle of a seasonally dry, sandy streambed. After undressing, I sat naked, cross-legged, in the sand. My gear arranged neatly by my side, within arm's reach.

Next I began to still my mind, so as to free it of mundane thoughts and distractions. The way I was taught to do this is to focus on what you are physically experiencing in the moment, which in my case was rather pleasant.

It had rained a few days prior, and the cool air of the morning was graced with the mildly sweet scent of desert flowers in bloom. It would get much warmer later on, but for now, the temperature was perfect, as were the small puffs of wind circling about in small eddies. It all felt so marvelous, even the cool soft sand of the riverbed beneath me.

Turning my attention to my other senses, I closed my eyes and began listening to the world. It was a magnificent chorus of sounds we seldom notice, like the sound of branches creaking in the wind. With all of my physical senses immersed in the experience, it happened.

In an instant, I found myself far out in space and so near to the Sun that its northern hemisphere nearly covered my entire field of view. Hanging there in space. I could sense the presence of others about me—wise, loving, and caring. They knew I was looking for answers and were helping me to do just that.

Suddenly, answers started coming straight toward me from the Sun, like beams of light. Each beam was much like the spine of a book; there was a title on each beam and so many were coming at me, I flew backward in space so that I could continue to look forward at them.

Many caught my eye, but I had to choose and I latched on to the ones I wanted most as the others zoomed by. After I read all the answers, the guides told me I could always return for more, as often as I wished. Then, they asked me to turn around.

It was an effortless spin and there, way below my feet, was the Earth, about the size of a schoolyard play ball. Likewise, I could make out a tenuous film of connection between my consciousness here in space and my body back on Earth.

The uplink message was pretty clear. "Come back. Things here are about to run amok." It was time, and yet I just hung there wanting to stay and not go back—to keep feeling that indescribable sense of love and connection.

Then the guides asked me to look. This is when I saw the most glorious sight of my entire life—a thick web of light and life, rooted in the major bodies of our solar system and stretching out into space far beyond the imaginable.

Within these webs, spirits floated freely through conduits of light. It was a blissful scene of sentience, love, acceptance, and connection. It was a feeling so profound that it is not possible on the material level. This is the very awareness that triggers your transformation. You know we're not alone in every fiber of your being.

There's no great unfathomable mystery about it, no need to plow through shelves of dusty, arcane texts or force yourself to pay attention through tedious and complicated sermons. We're connected and we're not alone. This is the Gnosis that completely erases any fear of death you may have, for the rest of your life. You know what you know, and nobody can take that from you. Never. It's your knowing.

Once I was given this knowing, the guides said that they love me and that they would always be there if I wished to return in the future. I sensed it would be a long time before I returned and that the offer was genuine.

After that, it was a swift and graceful return to the streambed. Like savoring an elegant dinner, I just sat there feeling the wind about me, listening to quail clucking, and contemplating what had just happened. It was a serene experience and, after it, my life would never be the same, for which I am grateful.

For those of you crossing the cusp, after you make it past The Great Winnowing, such experiences will be yours as you enter deep into the cusp. These experiences will transform you as well. You too will come to know what you know; you shall be filled with an unshakable resolve and will move with the flow to learn more as time goes by. Again, we're all a work in process, warts and all.

Of great importance is the fact that during these transformative experiences in enlightenment, you will come to find a sense of homeostasis—a stable condition that helps you to endure and to carry on with a new purpose in life. It will not make the going less difficult, but it will make the going more possible.

Here, we come to a point where it is time to make acquaintances.

Half a life after my first transformative experience, I began to write this book. This is when I turned to Rowena, my guide within The Elohim for a hint: "Where do I start with this chapter?"

Her answer was short and to the point: "Begin at the beginning," she said, "in Prescott." We'd come full circle. In the next series of 12 readings, you will hear Rowena speaking though me. These readings will give you the big picture in ways I never could, and you will hear warnings about what is to come and how we must fact it.

To me, there is nothing mystical or unknowable about it. These are messages from a friend who is further down the road and sending us information about what's beyond the next bend.

Above all else, never forgot. We have friends. We are not alone and we never will be.—*Marshall*

Reading No. 1 — Your Destiny

You are here because it is your destiny to lead others with wisdom love and compassion. It is a terrible burden to bear, as you well know, but given what is happening at this moment, it is one your world is telling you to accept with humility, integrity, and, above all else, commitment. Yours is a young species with great promise and potential. How you develop this and evolve is an ancient process for all life forms, wherever they may be in the universe. Now you must rise up or fall.

This is a process that is both painful and joyless for most. But for the few who endure, who stand in their knowing, who pursue life and the preservation of it with integrity based upon courage, there will be a new future after the smoke of the coming tribulation has cleared.

What brought you to this book, at this time, and in this way, is your own covenant with the Creator. The physical manifestations of what is about to happen to this planet - the violence and the ills - are all prelude to a greater tribulation yet to come. We are not telling you something you do not already know. But we may be telling you something you are still reluctant to accept.

For those of you with the integrity and courage to break free of your reluctance, this book will make a great deal of sense and will give you the knowledge to navigate the difficulties ahead. This is because the greatest tool you possess is your own the awareness. From it comes the knowledge and commitment to use that knowledge to survive.

It is less important to understand why a tsunami ravages a coastal city. It is less important to understand why a volcanic eruption inundates a city with a toxic hell. It is less important to understand why a tornado chooses to tear your neighbor's home apart while leaving your own untouched. It is less important for you to understand why new diseases shall ravage humankind.

All of this is less important than that which is paramount: your ability to use your awareness to lead you (and those who choose to follow you) away from these dangerous times and to make a difference.

There will come a time when your wondrous technologies and gadgets will become as useless as the clothes on your back. Perhaps just a few simple tools will be your new reality - the physical tools you as a species have used to survive in days long past. As time goes on, you will learn how to make better tools with your own hands, and they will serve you well as they served your forefathers.

All this is possible because the most important tool of survival is the magic within you. It is with this in mind that we ask you not to waste your time with tools made by others. Use those few fashioned with your own mind and your own hands.

This is not a new and unfamiliar talent, but one that has helped your species to endure far more of its brief history than any other. Therefore it is the goal of this book to help you revisit this most precious gift to you from the Creator. Wipe away the dust of time and polish your own magic so that you can see the future reflected in it. You shall go far.

Reading No. 2 — You Are Not Alone

We have been with you since the very beginning, as we have been with many other sentient races before you and those that are sure to follow.

Yours is a tribulation not dissimilar to those experienced by other sentient races. It is a culmination of time, fortune, and chance in the evolution of your species.

We have seen many fail and, if you succeed, as promising as your race may be, there are still no guarantees. You may nonetheless devolve and survive where other equally devolved species tragically failed and then lost themselves in the channels of time.

In the years ahead, your planet will undergo violent changes in response to other objects in orbit around your sun and from your own sun itself. There will be days of great suffering, pain, and death for those who are unprepared for the momentous times ahead.

Note, you have friends as well as enemies who would exploit your weaknesses for their own gain. These species squabble among themselves without realizing the cosmic need for consciousness, compassion, and cooperation.

The purpose of the teachings in this book - which were written eons ago - is to help you evolve into a more peaceful and enlightened species, enabling you to take your rightful place among the devolved and peaceful species of your galaxy. Your fate, should you fail this test of time, will be subjugation by malevolent races seeking to rape your world of its most precious resources: the magnificent diversity of life upon your planet.

The idea that you are being punished by a supernatural power is a faulty paradigm that you must dispose of so that you can evolve. Your species is not special. It is ordinary. It has been at the threshold of this evolutionary event before and failed because of corruption and greed. These same malevolent forces that have denied you rightful evolution in the past are once again working against your interests. They are from among you and other worlds.

Their hold on you is no greater than what you allow, if you are awake. A few of you are aware of this reality on the earth plane and your numbers are growing, as are the efforts to contain this natural and healthy growth in the global consciousness of the new and promising young species.

The desperation of their actions comes in anticipation of the great tribulation they know will soon come. They have known this for centuries and are prepared in ways most could not even imagine. Yet, even they know that, despite their tremendous efforts , there are no guarantees - for them nor for you. There is only one guarantee: your free will to choose your own path and with whom you shall walk.

These words may sound strange to you and unbelievable, but they are as true for you as they have been for other races in the past. The manner in which these natural tribulations happen varies from one world to the next, but all are in common with the effect: to evolve or not.

You need not rise up in great numbers to defeat these malevolent forces. Rather, you need only meet calamity in ways they cannot truly understand.

Their preparations are in the material world that is about to be shattered by a natural cosmic event. These events are a tribulation, the natural order of things in the universe due to its very nature and construction.

There is no mysticism or mystery about this. These events are, as they have always been, a reality of life throughout the universe. The goal is to weather them no differently than any other natural disaster.

By evolving through this coming tribulation, your species will understand the true nature of the universe, of its cycle of birth and destruction as a natural state of being created before all known life itself.

In a manner of speaking, it is the ship and what you say and do will navigate your outcome. You can either make it to a safe haven or run aground as result of your own shortsighted navigation. Therefore, it is most important that you accept responsibility as a species, to be a captains of your own fate. Leave this responsibility in the hands of others and they will misuse it to their own shortsighted benefit.

There is also encouraging news: other races are waiting to intervene on your behalf, provided you prove yourselves worthy. You must be ready, in sufficient numbers, to bring about the necessary evolution of your species so that you may continue unhindered in your development and so that others may embrace you as new members of the greater consciousness shared by the peaceful species of your galaxy and beyond.

Your world began to change decades ago and in ways that we and enlightened others can easily see and measure. Although most are oblivious to you, they are radically apparent to us. This, too, is the natural order of things. You sleep as your world changes so that it will awaken you with a violent shaking beyond your comprehension.

This is as it should be, for this is a test of the collective will of a promising young species. It is one destined, if it so chooses, to thrive in the years to come and elsewhere, beyond the reach of what you can now conceive.

We feel for your species a sense of excitement (as you would call it) for the true potential of the outcome. We see through the greater population of your species an opportunity for a greater number of involved individuals, through awareness of their gifts, to spark an evolution. It would be a spark so powerful and so great that it would present the honest opportunity to create the fertile possibilities for grand evolution, such has never benefited past generations.

Bringing these evolutionary advancements about is not difficult to understand, but achieving it requires a full commitment to your species and to its greater future. We stand with you who shall spark this coming evolution and advancement for you and your world, as do many others. We help because we wish to embrace you as enlightened brothers and sisters in what is a nourishing and marvelous cosmic consciousness that is now nearly within your reach.

As we have explained before and now happily explain again, it is about free will and how you choose to express it. No matter how difficult the coming years shall be, your commitment to the preservation of this gift shall always remain at the core of your ability to survive. We love you.

Reading No. 3 — The Tree of Life

You might ask, "What is this ongoing evolution of your species truly about?" It's a question that has been asked countless times on numerous worlds by sentient species you cannot fully imagine.

Life in the universe is so diverse that even we marvel at its diversity. Yet all life is connected, even though it is unaware of itself or the experience. Yet all life seeks to serve the same purpose: the expression of the Creator's design in which all play a necessary role in the growth of consciousness.

Consciousness means seeing what grows into mighty trees of life and reaches downward to fasten the soil of our material existence. To reach, to grow, to bind together is the purpose for which all life serves. In this regard, all life is connected like our species, which exists near the roots of the tree.

Those who exist at the tallest point are connected through the course of their brief lives to all those with whom they interact and are all joined in this greater community in the cosmic tree of life, to survive and coexist.

Humanity is joined by other races in the galaxy and beyond in much the same manner and it is reaching the point you call critical mass. That is, your numbers are sufficient to survive calamities that are a natural part of the process.

You may never know that all to which you are connected is not important. What is important is that you feel a sense of connection through this tree of life that encompasses all.

The trauma of what is to come will beset you no differently than it would the small creatures who dwell in the trees of your beautiful blue planet.

When you fell a tree, those native to it experience a sudden wrenching catastrophe in which many perish. Yet this is the natural order of things, for life to grow.

In this universe and beyond, comfort is always sought, but the lack of it is the natural order of things. That is what stimulates growth and survival and the path toward greater consciousness.

You will soon take a very big step toward the evolution of your own species so that humanity may survive and flourish - even though this and other hardships will prevail for a time and may seem unimaginable to you in this present reality.

In time, you will come to accept these realities, as difficult as they may be to embrace. When you have accomplished that, your path becomes clearer and easier to follow.

Yours is a beautiful world, but as all things will pass, it, too, will eventually fade over time. When that eventuality occurs, your species will scatter across the stars in numerous new homes of great beauty and resilience – but only if you prove worthy of this looming challenge to the survival and freedom of your species.

Believe that there is no such thing as death, but merely a change of state.

You must continuously take one small step at a time - even though you may never come to realize the full extent of the tree of life to which you are joined in the universe.

It is not necessary that you know. It is only necessary that you try, through small steps, to achieve awareness as you draw close to a journey that reaches beyond the horizons of your imagination. This is the majesty of it, that it appeals to your inner being to give you the strength to survive the difficult years ahead.

We help you as we hope you will help others in this microcosm of life, dwelling in the fibers of the tree of life, the very same tree of life that binds us all. In this way, the growth of your evolution will contribute to the vibrancy and health of the greater community in ways you may not see or possibly understand.

Your race is promising, offering a vibrancy that seldom occurs in this galaxy and beyond. Your species, as well as your planet, are like treasured pearls to be shared with all delight and wonder. But they can also be exploited and wasted by a few.

You will overcome these tribulations on your path if you stay true to the expectations of the Creator. You must try with every waking moment, with every breath you draw, no matter how tortured they become, in the years to follow.

Own this greater understanding, although it may leave you dissatisfied in the greater knowing of all the things. We experience this as well, the evolution and growth of life and consciousness throughout universe and beyond.

Strive each day to fulfill this purpose and it will replenish your inner resolve to take the next step. Try to come as close as you can to a greater extent that contributes itself to the betterment of all.

To try, to struggle, to move forward - even when hope seems buried under despair - is what the Creator expects of you. In these, you are not unlike ourselves. We, too, strive continuously to progress closer by understanding each element of the tree of life.

Only the Creator knows all and only the Creator sees all. Yield to the Creator and take unto yourself each small discovery as you grow and evolve. We love you.

Reading No. 4 — Fear and Ascension

The fate that is about to befall your world was destined eons ago and is unavoidable. What you will experience is a catastrophic interaction between your planet and your suns.

Your solar system, like many others in your galaxy, has multiple suns and you know them by many names. The second is the smaller, unborn, sun of your solar system with planets in orbit around it. This second sun is known to your scientists as Nemesis.

A term you use for one planet is "Nabiru," as it should be pronounced. It is one of the larger bodies in orbit around Nemesis, hence the name, which is the Planet of Crossing.

It is this planet in orbit around the Nemesis twin that will come closest to Earth. It will not strike your planet directly, but it will bring an entourage of smaller bodies that will be captured by the gravity of your planet and these will give you great misery and grief.

When you see it the first time, it will appear like a large comet in the sky with the tail longer than anything you have ever seen. As it approaches your sun, it will become more ominous and larger in the night sky. It is now approaching the south pole of your sun and will swing past it, back out into the depths of space as it has done before.

Your ancients witnessed this object with horror, as shall you. But you must be strong or it will defeat you first with your own fear. This is where those who have come before you have failed. They could neither contain nor master their fear.

This is why so many of your species are alive at this time, for this event of tribulation suffering and evolution. You are wiser now and less superstitious than your ancestors, because the knowledge of surviving this regular catastrophic occurrence is no longer hidden by a privileged few.

Readers of this book and others like it are experiencing awareness in much the same way that the elites and their priests in millennia past experienced it.

As was prophesied by your brethren many years ago, there is a web of life in your planet made of your own hands. In time, this primitive technology you call the Internet will open doors for you, doors to more powerful ways of sharing thoughts, ideas, and feelings with fellow human beings and other races in the universe as well.

Like a small creature breaking free of its shell, there comes the early light through the first cracks in what has always been an old and confined, but familiar, home.

For most, this change will evoke great fear of the unknown and that which lies beyond the comfortable confines of a long-familiar home, small as it may be.

Yet this cataclysm shall free you in ways you may not presently see (although a growing few of you are already sensing and exploring the first glimmers of light in your future). This is why we – and many others like us, in many forms, quite unlike your own – seek to help you with your own transition. We use the phrase "your own transition" because it is yours. It is not for us to control, or to sway the outcome, because only through your own efforts can this be achieved as we believe it can.

So when you see this terrible comet in your skies, do not let your fear grip your hearts so tightly that they cannot be undone. If you allow this to happen, the same fate that has befallen past generations on your planet will replay itself once again.

We cannot stress enough the great importance of your species assuming the burden of its own evolution at this time, for this shall be the last time when these objects return in the same manner.

Millennia from now, they will cause the final destruction of your world and the destabilization of all the objects in your solar system, causing a celestial reorganization as a result.

As the time approaches when these objects are closest to the center of your solar system, the skin of your earth will be pulled and reshape itself into a new orientation.

You call it a pole shift, but it is a very simple expression of what will be a turning point for your species. This is not an outcome you can negotiate using some naïve notion that you can placate an angry god, or even the Creator.

Rather, you must understand your role in the crucible of a major evolutionary event and prove yourselves worthy of Ascension. Ascension is not your right, but the privilege that results from achieving your goal as a species.

Those who will lead you in these trying days to come now walk among you. They are not kings and queens or heads of state, or even as the directors of mighty financial institutions, though you would logically believe them to be the ones. Those people shall fare as badly as most - but, perhaps, better than some as they are better prepared (at your expense).

They have known of these coming days for centuries and they have kept this dreaded knowledge to themselves. It has driven them to keep secrets, and to betray their own species through the presumption that they have the right to decide your outcomes, as though they are the only rightful co-Creators and you are not.

It is this misperception that will humble them when they and their children first gaze on this group of objects with their own eyes. Then they will know the insignificance of their powers and it will cause them to redouble their efforts in ways that will destroy *their own* chances for survival, more than yours.

This we see now, as we have seen before on many worlds elsewhere. It is the same story, told time and again for time immortal, for this is by design. The Creator, whom we serve no differently than you, set in motion this process of evolution at the outset of all creation. It is, by design, a test and must be accepted as such.

Embrace this with love and you shall achieve that which has eluded your ancestors: the peaceful evolution of your species to a new level of awareness. Here you will find spiritual bonds with one another that are new and fulfilling in more magnificent ways than you can now imagine and that extend to the greater community of life among the stars. We love you.

Reading No. 5 — Beware of Exploiters

Planning and preparation must be an urgent priority for all those seeking to survive and endure the years to come. This will be a time of tribulation that tests your soul and your resolve.

Are you committed to your species? This is the first question you must ask, for the answer will determine your spiritual strength to endure a tribulation of many years .

If you are in service to self, you may survive the early times of this cataclysm, but you will fall by the wayside as surely as the leaves from a mighty tree blown by the wind.

It is in this realization you must make your first step toward the future. It must not be a future of fear - although there will certainly be a great deal of that - but far beyond that, to a future of peaceful cooperation and mutual respect among all of your species.

By holding this vision of an enlightened future for your species, you will not only give yourself the strength to bear the tribulations, you will also escape them more easily and with less harm to you and your loved ones. This noble purpose does not drive you to the future, but rather pulls you, if only you let it.

Believe in this future and it will comfort you through the darkest of times within an inner sense of connection to what is most important for life on this planet. Your species will evolve, but you must avoid those who wish to tear you down and use you as you use draft animals to serve their own means and ends. Yes there are those species that, by making false promises, seek your demise through an abdication by you of your own free will.

Do not be lured in by them, for they are sweetly intriguing yet sit bitterly upon the soul once you have swallowed them instead of honoring your own free will.

This is the test for your species as it is the test of every enlightened species in the universe and beyond. You must embrace the sanctity of your own free will and that of others so that you live in harmony with all that is about you and within you.

The Creator does all things for a purpose, even that which you call evil. Put simply, it is the testing and reclamation function of the greater goal of life in your dimension as well as others.

It is the drawing power of evil that feeds on love to claim the weak. These are not the weak as you would think of them in a material sense, but the weak as you would know them in a spiritual sense.

In nature, predators are evil to the prey, but they serve a necessary purpose. They cull those who are weak and incapable of defending themselves due to infirmity, illness, or age.

It strengthens the herd or group when these predators remove weak individuals. All life forms, which are interconnected, are freed from the burden of breeding and nurturing those who consume resources better and served by the greater good.

Yes, this is a brutal truth and the reality of it can stymie the ambitions of a noble mind. Yet they are true - for you, and all other life forms. You are the givers of life and light but only if you choose to be what the Creator has made you capable of. If not, then luck and circumstance are your only friends and they can be as fleeting as a summer breeze.

We have seen your species fail this test of evolution many times before, but we now believe with the greatest possible confidence, that for the first time, humanity will do something it has never done before.

Your species will evolve and you shall be embraced by the many other species of your galaxy who have risen to the very same challenge. And they will know your pain and sacrifice as well as you.

In this, you shall find great compassion, brotherhood, and sisterhood, and above all else, a newfound sense of community within your selves and with those other enlightened species who await your decisive step forward.

These races know well that humanity will soon have the opportunity to learn through achievement. They know that, in the broad expanse of space, the distance between each of us is lengthened by fear and shortened through love.

Love is the most perfect expression of consciousness and it embodies within it the true intention of the Creator for each of us, regardless of which form we take or where we take it. Love, like the stars in a magnificent night sky, can light your way and give you the strength to do the impossible. Believe it. We love you.

Reading No. 6 — Hold Your Resolve

In the coming years, you shall see a growing trend in catastrophic earthquakes, floods, tornadoes, tsunamis, and more. They will occur in places you expect, but also in places where they have never happened before in the recorded history of civilization.

These tragic events will beset the earth like angry bees, swarming all living things near them. This is all part of a natural process that will build to a powerful crescendo of terror and fear. Let not your fears guide you through these events, but rather, your love.

A time will come when every man, woman, and child on the planet will know and understand that life as you have known it before these times, will soon end and that the world is shifting into new dimensions of existence and consciousness.

Whether humanity survives this process is the question each of you must ask yourself, for this is where the seed of consciousness reaches fruition.

Your technology will help save you, but it cannot work alone. To meet the challenges ahead, you must employ your technology and bind it with consciousness that stems from your love for others - not your love for yourself, nor your desires or greed.

Balance is the key in all that you do. In this sense, balance is not about dividing shares or counting seats in a survival setting. It is the balance of a fulcrum where you position yourself and your mind in a place where your need to survive, is equally balanced with the need of others to survive, especially those who seek you out for your wisdom and support.

The greater your need to serve others becomes, the greater your love becomes, in what you may call the synergistic balance of life.

As your planet begins to change and respond to new forces placed on it by the events to come, your fear and the fear of others will grow in proportion to the catastrophic changes you witness.

For many, this fear is avoidable yet they lack the understanding to avoid it, so they become dangerous and demanding – and, in some cases, demented by the effects of a sudden and terrible fear.

You will hear their wails and laments but you must not be moved by them nor should you serve them in any way large or small. Rather, gather unto yourself those who seek your awareness and understanding. Comfort them with your knowledge and compassion so they, too, may survive in service to others.

In the past, we have seen your ancestors face the same challenges that are soon to befall you. We witnessed with great sadness how they failed to hold their resolve, even though their consciousness was superior to your own in many ways. They failed not because their reasoning was flawed, but because their numbers were too few.

Those who knew and understood what was happening refused to share this knowledge with the masses, fearing a loss of their own control if they did so.

They understood very well what was happening, but held that knowledge in, serving only themselves. It is this way with all worlds like your own that encounter these catastrophic events of global change and the redistribution of life. In this, neither you nor those who withhold that which they know, are special or unique when compared to life elsewhere in your galaxy and beyond.

This is where you, the individual, make the difference and where the privileged ones perpetually fail. This is not to judge them for what they do best. This only fills the soul with anger that cannot be quenched and that masks the more essential truths that each and every one of you must know, believe, and embrace fully without reservation in all that you do.

It is difficult to understand these simple concepts in these times of plenty for the few. But many in your world live closer to these truths each day than the privileged among you. Yet they lack your mastery of communication and technology that transcends all of the limitations that have plagued your ancestors in millennia past when they, too, faced these challenges of evolution.

Do not repeat these past mistakes because you choose to do so, because you choose to ignore the knowledge carried forward to you, because you choose convenience in serving yourself and your own wants and desires. There are no rewards for rising above feigned self-interest other than those that you bequeath to yourself in this life and beyond, for life is eternal and that which you do today is with you and all your tomorrows.

Judge yourself now, for your wisdom, courage, and insight. Then share that with others with love and compassion as circumstances bring you together. When sentient life forms share in this way, they multiply their numbers with the strength needed to survive what many will come to view as the survival. Know this. We love you.

Reading No. 7 — Nemesis, Planet X and Our Sun

Your world is already experiencing changes in advance of the bigger events to come. The signs are there for those in awareness. The rest will find awareness as the major events begin to happen with more severity and greater frequency.

As the object you call Planet X, and all the many other names given it by your ancestors, approaches the central perimeter of the inter system, the events you now experience will become greatly magnified. This is because of how this object and its constellation of planets and moons is interacting with every other object in your solar system, especially your sun.

The greatest threat you face in the tribulation will be from violent storms on your sun, as it wrestles with the magnetic energy and gravitational forces of its smaller twin, the Nemesis star as some of you have called it.

The greatest peril your planet faces will be from the heat and radiation caused by these interactions between your sun and its twin.

We have watched from afar this inevitable cycle of destruction and rebirth of your world and what that means for life upon it. As beautiful as your planet is, it has a violent fate cast by an unnatural alignment of stars in your system.

It is why life on your planet evolves than devolves and disappears, and why so few species endure the test of time and travail.

You will also see meteors in your skies rain down on the surface of your planet as though part of great storms. They will destroy great swaths of land leaving not one building, not one home, to stand as before.

The frequency of these storms will cause great fear because they will happen night and day and without warning. Yet these, as deadly as they may be, will pale in comparison to the horrors of a monstrous solar storm. Again, we warn you of the heat and radiation that will poison your bodies and your lands and make your water unfit to drink.

Desperate people will steal from others and commit terrible violence, to take what is yours and what belongs to others. They are only interested in themselves and can only destroy what was created by others.

These lost people will not endure the test of time – and never do. So do not let them feed on you and those you love. Avoid them and let them exhaust their own resources so that they will begin to feed on themselves and, in time, the last of them will perish, alone and in fear.

You must therefore avoid these angry, selfish people, these lost souls who turn away from the light of the Creator. Their fate is already sealed but yours is not fully written. Follow your fears and you will share their most definite fate with them, at their hands.

They will plead with you for help and decency, but you must show them none, keeping your distance from them by standing in your integrity. Rather, be strong and be kind to those who seek to help you and others to survive.

It is the terrible burden of evolution to see those among you who have lost their way as they find a final destination of their own making. Be strong and love those who are not lost, who seek the natural path of evolution in the light of the Creator.

There are no guarantees and, while you may not see the future that others will, they will be worthy of it. Believe it. We love you.

Reading No. 8 — Signs of the Pole Shift

In the coming time you regard as a pole shift, the early warning signs will come from the infrastructure in your cities: bridges, dams, and so forth. These artificial creations were not intended to handle the profound changes in their environments that are already underway.

You must anticipate these failures and, when possible, avoid exposing yourself to the consequences of sudden, unforeseen, and catastrophic failures.

These will happen with great frequency and severity in the months and years to come. Once the danger has passed, very little of this infrastructure will be serviceable or useful.

Natural caves offer the greatest level of protection, unless they are flooded. Avoid all subterranean refuge that is adjacent to large bodies of water, which will be in motion with terrible effect in the times to come.

The shelter of natural caves that are far away from large bodies of water will give you the safest havens from the heat and radiation, as well as the movements in the outer skin of your planet.

There is a time to seek these places of safety and refuge and that time is now. Seek out these places with your friends, family, and children and all those whom you love and cherish. Then, prepare to take refuge there when you believe the time has come to move to safety.

The privileged classes and government leaders have already been doing this for decades in preparation for what they know is to come.

Their ambitions are born of their own fears, their fear of death, and - of equal importance to them - their fear of losing control over the masses they govern. Their fate is not as assured as they seem to believe, for their preparations are born of their need to control.

Your need must stem from a need to coexist with others in harmony and peace, no matter how difficult the times become - and they shall become terribly difficult.

This disaster to come is very bad, but it is not the worst that could happen. You can take some comfort knowing that the worst that could happen, is not what we foresee for you. Not this time.

Other generations to come hundreds of years from now will face the same challenges as you will, but their fate will be many times more severe.

This is why your survival in this cataclysmic event is so vital to the continuation of your species. If these future generations are no better prepared than yours, they, too, will fail. Then, the last chapter of human history will be written in the ancient dust of time past.

You must therefore not only live for today and for one another, you must also live, rebuild, and prosper for the sake of your own species' ability to continue.

Other species elsewhere in your galaxy have, like your own, reached this point of decision and continuity - and they have failed. They did try to overcome their own adversities. However, by simple chance, they were unprepared by those who passed before them to deal with their own fears and to believe in their own abilities as a unified species to exist beyond calamities.

The most important thing you must remember in all of this is that, without a committed love for your own species and a certain belief in its continuation, people will turn against one another and against yourself, in most unfortunate ways.

Focus on your fears and you will not understand the natural order of life in the universe, that being survival. It is not easy and the punishments are severe when species are divided and quarrelsome.

This will be the greatest lesson of the coming pole shift. It will be a long, troubling process of many months punctuated by brief moments that cause terror for those who cannot rise above their own fears through a genuine love for others.

As your Earth's orientation and the appearance of its outer skin change in the coming pole shift, the greatest change can come from within the souls of humankind and the inner core of what motivates and guides them to wisdom.

This will be a time of great tribulation, yet simultaneously a time of magnificent empowerment for the promise of a golden age in an enlightened future for your children's children and beyond. Grasp this opportunity for what it is: a coming decisive moment in the evolution of your species.

We believe in you and, as young as you are, we believe your species will find the wisdom in each and every one of you to reshape your future even as your world reshapes itself. We love you.

Reading No. 9 — Sharing with Others

In the days and months to come, you will begin to see a progression of catastrophic events - both natural and manmade. Both are caused by the same process and will exacerbate each other in ways that cannot yet be foreseen.

Your species has learned how to force its will on a natural system of checks and balances and, consequently, much of the suffering will be heightened by this combination of factors.

The seas of your world will become your greatest threat, as forces greater than you can imagine are unleashed in a natural, progressive cataclysm.

Coastal cities around the globe will suffer the most as sea levels rise and great storms and tsunamis devastate many coastal regions. People there will need your greatest compassion and help, for they are the least prepared and shall become the most needy.

As events unfold, do not judge these people for you would likely make the same errors if you were in their place. For this reason, there is no right to judge others who suffer calamities. Embrace them with compassion and strength.

They will flock to your sanctums of safety and, when they do, share what you have without the expectation of reward. This is for two reasons. First, judgments always clouds reason and when you judge others, reason is the first to fail. This is because judgment is born of anger and arrogance and these emotions will serve you badly if you allow yourself to succumb to them.

Create your own stockpiles of basic foods and medicines now while they are plentiful and affordable within reason. You need not entertain your survival guests. You merely need to help them with subsistence and basic shelter.

They, in turn, will strive to organize themselves and many will find the strength to participate in the greater survival of the whole. In this, they will bring valuable new skills and knowledge to you and all those who choose to help.

The children among them will have the greatest need and this is important because they are your greatest hope. In the cataclysm, the resilience that is so natural for a young child will inspire you to take one more tired step. To shoulder the heavy burden of your responsibility, knowing it has replaced a life that was somewhat stable.

They will intuitively understand the gravity of the situation and will adjust to it rapidly. They will find ways to contribute and be a part of the greater effort of survival, if they are old enough.

The innocents among you will be your greatest treasures. Nurture them first and help their mothers and fathers care for them when they are too exhausted to rise up for even the most common tasks.

Find your commitment to shelter and nurture the youngest among you and you will find a great purpose that sustains and nourishes your spirit in the darkest of times. The children among you are a magnificent gift of joy and purpose for a greater aim: the survival of your species.

Expressing your love through service to others is also the best way to survive. Contention born of self-interest steals away opportunities for survival for all those who succumb to their own self-interested weaknesses.

Such people can never inspire a child to wipe away tears and continue on, nor can they lead with true compassion. Avoid these people - not because you perceive evil in their thoughts and deeds, but because they are less likely to survive the years of tribulation, which are certain consequences of what is to come. We love you.

Reading No. 10 — Heat and Radiation

The pole shift, as you call it, is the natural realignment of the outer skin of your planet in response to external and internal forces. It is a natural process of devolution and change and is experienced throughout their existence by many planets such as yours that possess a molten core.

The principal cause of the coming pole shift, as you refer to, it will be the internal forces of your own planet in response to the external influence of your sun. Great solar storms will warm the surface of your planet and these forces will penetrate deep into its core.

This additional heat and radiation will exacerbate the natural processes within the core of your own planet, causing it to "boil up," so to speak, much like a soup kettle placed closer to the flame.

These bubbles of magma will exert forces on the outer skin of your planet that are far greater than to which it is accustomed. The eventual pole shift will increase the natural speed and inclination of existing and ongoing earth changes. These processes are what will propel the shift event itself. There may also be external influences from other bodies in your solar system, but they will be minimal compared to the principal causality, that being your sun.

Safety for you, your loved ones, and those whom you have embraced will then depend more on your ability to find safe refuge during the drama of these cataclysmic changes.

You must learn to become creatures of the night as you move about and forage for food and water.

During the days, look for the deepest natural underground shelters nearest you. Ideally, you want caves or caverns that have endured previous cataclysms in the past, especially those with their own supply of natural artesian water.

Water, more than food, will be your greatest concern. Your frail bodies cannot endure long without this vital replenishment to your vital juices. Filter it carefully and treat it for bacteria and poisons as these will be even more of a threat to old, malnourished, and weakened bodies.

You must also teach yourselves to grow nourishment indoors, away from the heat and radiation of the day.

Your skies will no longer be blue. But, in time, that will return and with greater clarity and magnificence than you could presently hope to see.

Until then, cover your body completely in linens and other lightweight natural materials when you are walking about in the daylight. Any color will work, but white is best or, alternatively, pastel shades. Fasten these linens and sheets loosely so that your skin will be more comfortable and easier to protect.

Your skies will appear sullen and slow-moving with dark patches that will cause sores on your bodies. They will heal slowly or in some cases not at all.

Cleanse your skin with simple dirt or clay if water is unavailable in sufficient enough quantities for you to wash. It is better to be dirty with simple soil or dirt than to allow such deposits to remain on your skin.

Even if you are exhausted past the point of caring, you must tend to your own bodies or they will succumb quickly to the harmful gases and chemicals that will infuse your skies.

There will be a slight darkening of the daytime sky, as a swirl of red-streaked darkness encompasses the whole of the earth. There will be no blue sky anywhere on your planet during the worst of this cataclysm, so there is no escaping to another place or another land. Rather, accept it as it is, where it is, and know that there is no better elsewhere (except, perhaps, underground).

You will adapt over time and find new ways to survive so that your species can carry on. Have faith in this and your own abilities to adapt to changing circumstances, for this is the single greatest talent of your species: its innate ability to adapt to cataclysms that will certainly destroy a majority of other species on the face of the earth. It is part of what makes you worthy as a species: your tenacious ability to band together and overcome the difficulties you shall encounter on your path of evolution as a noble and promising species. We love you.

Reading No. 11 — Consciousness

We wish to tell you now of an extraterrestrial visitation that will occur in the midst of the cataclysm and beyond.

Your world and your species are not alone. You have never been alone and, in time, you will understand that life in the universe is widely scattered throughout the stars.

Much of space is like your great deserts, a place where life in its smallest forms struggles and where sentient beings travel far to reach distant locations where resources and sustenance can be found.

Your greatest treasure is your diversity of life and its many magnificent forms, more so than on most worlds. You must know this, for your world is both an object of admiration by many, as well as an object of exploitation for others.

Your struggle will be won not by fighting the exploiters of your world, but by reaching out to those races eager to embrace humanity as a new and spiritually enlightened species.

Those who approach you with commandments and solutions only they may provide, are not your friends nor have they ever been your friends.

Their aim is to delude you with your own contentious fears into accepting what appears to be the greater power of their technology.

They do this because they seek dominion over your world and its many resources, which are of great value to them. They *take* as opposed to *share*, because it is their way. Your way, if it is to prevail, must follow a different path.

You must find the will and wisdom to contact those species who respect your free will as well as their own. They will help you, if you are on the path of enlightenment and consciousness, in oneness with them and yourselves.

If you fail this test of evolution through your own fears, hates, and greed, you will place your destiny squarely in their hands, which is exactly what they seek. They want a completely free hand to do with you and your worldm, whatever serves their own interests, above all others.

Life in this universe is a precious thing, but its continuation in its many forms is not guaranteed by the abuse of others, but through the path each chooses as they journey through the lessons of wisdom toward the Creator. The consciousness that created this universe in a single thought, as it has others.

Fear is the most important thing to avoid. Stand in your integrity and, through awareness, act on your own self-determination to be a sentient species with a beautiful world.

Resist all those who would enslave you for the resources of your world through trickery and beguiling language, designed to manipulate your fears, expectations, and judgments of others. All of these negative emotions are the handiwork of those who steal from others by enlisting their unwitting help.

Other species have faced the same challenge and have lost their worlds time and again. Be more like those who have risen to the challenge with enough courage to love their worlds and protect their own right of self-determination, no matter what the cost in lives or treasure may be.

Self-determination is it essential if you are to walk in the path of the Creator. All that you are about to suffer as a species will either draw you closer to your path toward the Creator or enslave you for the benefit of others.

This is your greatest test, more so than the actual test of survival itself. Species come and go with these tests, whether it is their deserved fate for their own weaknesses or not.

View these coming years of tribulation and suffering not as a mindless, random penalty, but as an opportunity as a species to claim your place among the stars alongside other races that, having endured similar pains, chose the path of the Creator.

These few other races that have overcome adversity and now enjoy peaceful existence wish to welcome you into a greater community of evolved life, where coexistence is the principal goal.

In the coming years, you will learn the lessons of your long-ago past, forgotten in the shifting sands of time, yet always there. There are no such things as "super species" with intellects beyond your own understanding.

Rather they possess the benefit of time and experience, a time much greater than yours but no less flawed or imperfect. They are simply motivated by their own wants, needs, and desires and will leverage their advantage to achieve their own aims, knowing their efforts are far from certain, should your species resist.

There will come teachers to show you this simple form of communication with enlightened species who are listening, quite assiduously. They can and will, with great happiness, help your species understand what they have discovered in their own journey to the Creator.

Your species is at the threshold of its most important next step toward the path that will lead you to the Creator as a whole species and as individuals within that species.

Succeed in this effort and many others will greet you with open arms. They are here on this earth plane to assist you in peaceful and loving ways and to nurture your own leadership and self-determination as a species.

Only then will you find them and their embrace. Fail this and you will remain in the isolation your species has always known and regretted, here and there along the way.

Your attempts to communicate through radio waves sent by other species, miss the expanse of time between the evolution of other species and their technologies, as compared with your own. Just as smoke signals were ones he used to communicate across distances, radio as you use it is one similar step.

The ability to for enlightened species to communicate multi-dimensionally is very different from your present day modes of communication. They have learned to transcend the limitations of material technologies such as radios and microwaves.

These technologies, although powerful across relatively short distances, cannot span the depths of space that separate you from other species. It is why they communicate through consciousness - for it is immediate and complete.

In time, your species will adapt and, hopefully, evolve to a point where you understand your own internal abilities. When you do, communication will become effortless. Only the gifted few among you can now achieve with some measure what will become commonplace for your species.

Then, you will transcend the petty fears, greed, and deception that have always plagued the development of your own consciousness.

When you have mastered the basic strengths of your embedded abilities, they will flower for you in ways you cannot possibly imagine today.

Your tomorrow holds great promise for you in consciousness, cooperation, and joy. May you all live to see it in your lifetimes and experience the grander majesty of consciousness through the Creator. We love you.

Reading No. 12 — The Purpose of It All

This last reading in this series of teachings is to explain to you how we all benefit when a species evolves into consciousness at an enlightened level.

There is a natural order for all things in this universe and beyond. We, along with everything else, are the genesis of a single thought, a need felt and expressed by the Creator in a rudimentary physical form.

When a species survives the test of time and evolution, it becomes a vehicle for the expression of the Creator's will to enable a desired outcome: to join the Creator in the creation of consciousness throughout the universe and beyond.

Yours is a younger sun, approaching the prime of its existence. All material things material follow a similar cycle of creation and destruction and then rebirth in a new form, whereas consciousness and its highest states endure beyond this cycle in a material world.

There are vast areas of space larger than your minds can truly comprehend. This void is like the great deserts of your world, where life is sparse and lowly evolved. Yet just as a desert may bloom in the presence of water, this vastness of space is like a dry and unfertile soil that awakens to a new purpose, with the moisture of consciousness.

Beyond our abilities to perceive this vastness, lies the greater intention of the Creator and we are drawn to it no more strongly or no less strongly than your own species.

We are simply on different points along the same path, replete with treacherous twists and turns leading to green pastures that momentarily comfort the weary who travel the path to the Creator.

Your planet earth is one such a oasis of green pastures and cool waters along a trail strewn with the failures of those who have gone before you. There are those who have reached out for consciousness, but through their own failings were unwilling to reach far enough and in so doing, now dwell in the material, knowing their eventual fate as a species is sealed.

Your human bodies are mere adaptations of life in a narrow slice of time so small it easily goes unnoticed. As a species, you will evolve into consciousness and, as you do, your bodies will adapt to new skills and talents. You already possess them deep within the very code that orders the regulation of your own bodies.

Your ancient forefathers were as different to you today, as you shall be to your decedents in millennia to come, if you are successful in your own evolution.

As we have said before, this is not the first time your species has reached this all-important point of evolutionary decision. You have been here before and have always failed before, but we believe this is your time.

This is your time to transcend your difficulties and shortcomings, whatever they may be, as a unified species devoted to consciousness and peaceful coexistence.

Even then, assuming you have achieved that which we believe is now possible, there will continue to be new trials and tribulations for your species, new threats and hurdles to overcome. These you will find in time and, as difficult as they may be, your future travails will not be as fraught with the danger of failure as what your species now faces.

This is your clarion call. Accept the weight of the responsibility to yourself and to your species for this grand purpose of evolution, conceived by the Creator and served by us all.

We, ourselves, are what you would call energetic beings, entities who have left physical form for a more pure state. Like you, we seek to join the Creator. Like you, we cannot do it without you and others like you. It is why your species is so vitally important to us.

Like all such promising young species as your own, there comes the opportunity for all of us to walk that path of the Creator together and draw closer to the Creator together.

Wherever you may be along the path of this journey, that is simply where you happen to be. It does not diminish your achievements, nor does it rank your abilities for all that truly matters.

The only measure of worthiness that applies fairly and equally to both you and us and all other conscious-enlightened beings, is that we choose to progress.

In doing so, our energies combine and nourish one another in the most marvelous ways imaginable. In the fullness of time, you will come to understand this and it will please you and satisfy your aims to become more than you presently are. This is your next step on the path toward the Creator. We await you and we love you.

Part 3
The Good News

OTHER BOOKS BY MARSHALL MASTERS
Planet X Forecast and 2012 Survival Guide
Godschild Covenant: Return of Nibiru
2012 Wisdom of The Elohim
Indigo – E.T. Connection

10

We Can Do This

For more than a decade, I have been researching and writing on the topic of Catastrophisim. In all those years, the usual questions I've been asked are:

- What should I buy?
- Where should I go?
- When should I start?
- Who do I do this with?

While these are important and worthy questions, they don't include the most important question of all: What do I already possess within myself to survive? Above all others, this question is the transcendent one, and yet it so easily languishes in the modern mind.

This is a consumer response; when we first become aware of a problem or issue that troubles us, we're drawn back to our indoctrination unconsciously. This is because the indoctrination of consumerism methodically programs us to see ourselves as incomplete. Consequently, we feel a synthetic need to acquire a plug-and-play answer that calms our concerns with the bliss of consumption, as opposed to seeing ourselves as already complete persons.

Although this indoctrination is subtle but powerful, people are nonetheless becoming aware of it, mostly as the result of the present global economic morass. However, whatever is driving this change is not as important as the change itself. For example, one sign of change is that more people are realizing that when it comes to using credit cards, it's better to leave home without them.

Yet, residual aspects of this deeply ingrained indoctrination of consumerism continue to carry over into our personal preparation efforts for the difficult times ahead. A good example is the companies that offer the convenience of packaged survival kits.

The Bucket List

A true single-source disaster solution, bucket kits are perfectly suited to our modern lifestyle. The acme of survival ease, these kits are typically offered in conveniently labeled buckets—each with its own value pricing list of the contents. The appeal is obvious. In case of disaster, just add water.

Consequently, where the buckets are stored becomes the most pressing issue for catastrophe consumers. Most just add storage shelves in the garage or basement for their buckets. Once the buckets are neatly stored away, preparation for 2012 is complete and it's back to business as usual (i.e., time to go shopping again).

Americans didn't always think this way. There was a time in America, prior to industrialized agriculture, when many small stores rented out canning rooms in the back. Families could harvest fruits and vegetables from their victory gardens and use these canning rooms to preserve their own supply of jellies, jams, pickles, and so forth for the cold winter months.

For merchants, the sale of canning jars, lids, pectin, and related supplies was a good source of income, proving that self-reliance is a viable business model. Little stores with deep roots in the community, as opposed to national retail chains with aisles burgeoning with products made in China, could survive. Consequently, globalism has resulted not only in outsourcing jobs we need to raise families and pay mortgages, it has also drained us of our strong sense of self-reliance.

Without a strong sense of self-reliance among the people, America would never have grown to the size it is today. Without the courageous settlers who loaded their possessions and families into Conestoga wagons and headed west into the unknown for a better life, where would we be today?

Early settlers did not buy Go West buckets. They selected and packed everything they took with them; they knew exactly what they had and how to use and repair it. This demonstrates why the consumer-driven market solution of survival buckets is counterintuitive to the very goal of survival.

When you buy a stack of survival buckets, they come sealed. Each bucket comes with a convenient list detailing the various products inside, and each bucket contains several different products. However, the people who pack these buckets always leave out the most essential feature: the thought processes that go into packing the bucket itself. Ergo, the problem is that you do not receive this thinking process—just a bucket and a bucket list.

Now, picture yourself in a future time, in the aftermath of the disaster. See yourself frantically reading through bucket lists to determine which buckets to unseal first. As you do, know this: Whatever is in that bucket had to satisfy three core business requirements:

1. Whatever is in the bucket has to look impressive on a bucket list.
2. Whatever is in the bucket must be profitable for those who put it there.
3. Whatever is in the bucket must offer a generic one-size-fits-all solution.

Now ask yourself this: If you're fighting for your life after some dark future catastrophe, is this the best time to read through a bucket list for a description of whatever is in your one-size-fits-all bucket solution?

Will Mother Nature put the calamity on hold so that you can put a nice clean tarp on the floor of your garage, lay out the contents of your buckets, and use the lists to inventory your just-in-time solution?

Or, imagine you have just survived a cataclysmic event and you are far away from your garage and your neatly stacked survival buckets filled with all kinds of useful (one-size-fits-all) products. Now you're miles away from your buckets, and between you and your garage are collapsed bridges and overpasses, broken mains, gridlocked streets, and impassable highways. Odds are your neighbors will find your buckets before you do.

The point here is that consumer-driven bucket-list thinking is like enrolling in a new drug study and learning that your name is on the placebo list. So, are you ready to toss this bucket-list placebo consumer thinking for something that works?

If so, then resolve to pack your own Conestoga wagon, and get started today with the transcendent question obedient consumers never ask: What do I already possess within myself to survive?

Deep Blue

As consumers, we're indoctrinated to leave such thinking to the people selling us neatly packaged, completely tested solutions. We experience zero learning curve and we can return our buckets within 30 days of purchase or before the cataclysm, whichever comes first.

This is where experimentation, evaluation, and trial-and-error are our first and best tools for increasing our odds of survival, no matter what comes. Furthermore, the more time and thought you put into these tools, the better they will serve you.

A brilliant example of the heuristic process is the story of how IBM's Deep Blue supercomputer beat World Chess Champion Garry Kasparov in 1997. Kasparov was a formidable opponent—a man who would be ranked 25th in *The Daily Telegraph's* list of 100 greatest living geniuses in 2007.

In 1996, IBM's Deep Blue first played Kasparov, and Kasparov won the match 4-2. Nonetheless, Deep Blue showed that it could win against a human opponent, and it was extensively upgraded for a 1997 rematch with Kasparov. As a result, Deep Blue became the first computer to defeat a reigning world chess champion.

While Deep Blue got the limelight for beating a chess champion, this victory was not about a machine beating a man. It was about a large team of people who built a machine for a single purpose: to defeat a champion like Kasparov based on their analysis of the game of chess. How did they do it?

After loosing the first match in 1996, IBM hardware and software engineers had something they'd never possessed before a prototype experiment with a treasure trove of useful data, which they then used to design and build a second-generation Deep Blue.

Whereas the first-generation Deep Blue had been designed to defeat a chess champion, the second generation was specifically designed to beat Kasparov, and it did, in a very close match. Therefore, Kasparov was in fact defeated by a group of humans jointly experiencing a heuristic process.

The point here is that for these engineers to create a second-generation Deep Blue capable of beating Kasparov at chess, they first had to visualize that goal in their own minds, thereby building new neural pathways within their own minds.

In a very real sense, the Deep Blue engineers used their own brains as self-programming wetware computers. So, no matter how historians slice it or dice it, the seeds of Kasparov's defeat were sown in the human soil of a heuristic learning process.

The lesson here is: If you want to live, never forget. Bucket lists are about dying so do your own survival homework. Follow a goal-driven process where visualization is your chalkboard and expect to learn from your mistakes, after you take responsibility for them.

Visualize the Goal

When gathering our own survival supplies, conventional consumer-driven indoctrination tells us to begin with bucket lists. To do this, we rely on shopper tools: product comparison charts, sales notices, discount coupons, reviews on the Internet, and so forth. These tools tell us about a particular product, relative to competitive offerings.

In other words, these tools are focused on the present to help consumers get the best value. They are not about using the product in an unforeseen future. Ergo, how can these tools tell you that the products are appropriate for the future, unless you're willing to accept throw-down assumptions for the promise of a hot deal?

With this in mind, visualize yourself in a not-too-distant time, and you have just survived an unexpected cataclysmic event miles from your home and all your useful products. At this point, the only bucket you will have is the one attached to the top of your brain stem.

Whatever will be in that wetware bucket on your head is going to be one of two things: pre-programmed neural pathways ready to go to work for you or a synergistic vacuum of global trading potentials in a just-in-time, plug-and-play, automated solution opportunity pool that you'll have to fill as you go along.

Ergo, use the visualization of future events to set your mind in motion. Your mind will then create the neural pathways needed to help you survive the kinds of events you're visualizing.

Unfortunately, this is not what most people visualize at first.

They see themselves joining others, beating on the gates of the White House, and screaming, "OMG! Why do governments and elites get to survive in underground arks and callously abandon us to the surface to suffer and die! We've been betrayed and we're mad as hell." Sounds great, but after a lot of huff huff, puff puff, and pant pant ... you're still at OMG!

Now turn around and look at this from another perspective. What really is the best way to situate yourself for the tribulation? Will you survive in place or on the move?

Or, in simpler terms, will you be a bunker bunny or an Earth Pilgrim?

Bunker Bunnies and Earth Pilgrims

A bunker bunny plans to hunkerdown through the lean years in an ark, underground or at sea. This is why survival firms are selling golden Wonka tickets to those who can afford them to reserve a place in their arks.

For those with pockets deep enough, this is a very attractive consumer-driven solution to surviving a global tribulation. Consequently, the process is essentially like booking a cruise, except that all you can get is an inside cabin below the waterline.

Once you're in the shelter, you're like a bunny rabbit in a burrow because you run the same risks as those above and you know at some point you will have to emerge from your shelter to find what you can only imagine. In the meantime, you wonder whether a large predator will discover your hideout and slither in to get you or dig you out like a dog would a bone. Face it, like any bunny in any burrow, you'll sit, wait, and worry.

Although your physical needs may be met in your shelter, survivor guilt will be tremendous. These feelings of survival guilt will be especially hazardous because treasured friends and extended family members will be left behind.

During The Great Winnowing, these fears and regrets will spawn fear-based emotions. And as with all fear-based emotions, they will become a slippery slope toward death, despite physical comforts those left on the surface could only imagine.

In contrast, the vast majority of those who do survive the cataclysms will do so as Earth Pilgrims, a term first coined by Solar Code author Echan Deravy for those who will be abandoned to the surface. Earth Pilgrims see the dark cloud of fate with a distinct silver lining.

The bunker bunnies will be doing something that has been unnatural for our species; they will be surviving in place. Those abandoned to the surface will be doing something that has always been natural for our species, surviving on the go.

But haven't we evolved into a survival-in-place species? If that were so, we wouldn't gain weight so easily. Our bodies are still programmed by nature for a survival-on-the-go, hunter-gatherer existence. We are survival-on-the-go beasties, so accept no substitutes.

One way to help us understand the major difference between survival in place and survival on the go can be found in another species with a mighty legacy of surviving global cataclysms. This is the Coast Redwood, and how the species has survived is an amazing feat of nature.

The Coast Redwood is an ancient conifer species that presently ranges from southern Oregon to central California, extending no further than 50 miles inland. These giant trees grow to more than 300 feet (91 m) in height, and the oldest on record dates back to the days of Jesus.

Of note is that, geographically speaking, the Coast Redwoods were not always so confined. At the time of the Cretaceous–Tertiary extinction event, approximately 65.5 million years ago, Redwoods were the dominant species throughout the Northern Hemisphere. At that time, the Earth was warmer and this sub-tropical species thrived.

However, when a massive asteroid struck the Earth offshore from the present city of Chicxulub in Mexico's Yucatán peninsula, the dinosaurs died, but the Redwoods endured. Why? Because Redwoods are perfectly adapted for survival in place; they possess genetic variability several times greater than humans.

Modern humans are diploids, with only two sets of chromosomes stored on 23 chromosome pairs. A Coast Redwood, on the other hand, is a hexaploid—each of its cells contains six sets of chromosomes, for a total of 66 chromosomes.

Redwoods are also a collective survival species. Their roots are surprisingly shallow and can extend 100 feet (30 m) outward from the trunk. Plus, they merge with the roots of other Redwoods so as to share water, nutrients, and strength.

Consequently, if you were standing in an old-growth Redwood grove during an earthquake, you would not shake side-to-side as you would in a modern city. Rather, you would bounce up and down as though you were on a trampoline. Ergo, your chances of surviving a major earthquake inside a Redwood grove with 300-foot (90 m) trees looming about you are actually better than survival in most manmade structures.

This is why any firm selling berths in its survival bunker is in fact offering an unproven solution, especially when you compare the solution to a species possessing a solid track record for surviving global cataclysms.

If you are committed to a survival-in-place strategy, whoever designs that place will make all the important decisions for you, so choose these people wisely.

For those of you committed by choice or otherwise to a survival-on-the-move strategy, your preparations must begin with what nature gives us. In other words, what do our 23 chromosome pairs buy us?

Nature Designed Us to Be Earth Pilgrims

The reason why Redwoods need 66 chromosomes to survive is that they must survive in place, wherever they're rooted. They cannot walk the Earth as we can, which is why we have 23 chromosomes. It's all that's needed for a highly intelligent and adaptable species designed by nature to walk the Earth.

If nature designed Redwoods and humans the same way we buy airline tickets, Redwoods would get socked with extra baggage fees, whereas all we would need is a small carry-on (or just a finger toothbrush for that matter) and we're good to go. Ergo, less is more, providing we use what we've got.

As an Earth Pilgrim, your ability to survive is built in by the Creator; no extra genetic baggage fees or trusting your life to untested technologies is required. We all pop into the world ready to go the day we're born, but that's a fuzzy distinction for naysayers. Rather, naysayers are quick to point out that our confidence is naively misplaced or overstated, and as proof of that they'll use what happened to the Neanderthals. But is this comparing apples to apples or apples to oranges?

When you compare Neanderthals with Cro-Magnon, the ancestor of modern man, the Neanderthals had the hands-down genetic advantage. Their brains were larger and they were arguably more intelligent because they were the first to bury their dead, and they could make better stone tools than Cro-Magnon.

Furthermore, Neanderthals were better adapted to harsh climates and were stronger, and the many mended fractures in their bones tell us they could absorb brutal levels of physical punishment. Hence, the naysayers, like Las Vegas odds makers, would bet on Neanderthals to triumph in the last struggle for dominance. Really?

Naysayers are lazy-logic people with a personal disinterest in doing their homework. One does not become a successful Las Vegas odds maker by thinking like a naysayer.

Rather, your odds makers do their homework and scalp the naysayers more often than not. So, in this case, what would a savvy Las Vegas odds maker see when figuring the percentages in a final showdown for control of human destiny?

For starters, he or she would see that Neanderthals and Cro-Magnon represent very different branches of the same species, though present scientific findings seem to indicate that the genetic makeup of some of us living today includes traces of Neanderthal ancestry.

However, the odds maker will not waste time debating missing-link solutions to explain the differences between Cro-Magnon and modern man. Rather, they'll look for that wicked

Cro-Magnon right hook that seems to have singlehandedly neutralized the Neanderthal's considerable advantages.

In a word, that deciding difference involved where nature happened to place the larynx and the structure of the hyoid, a small bone that holds the root of the tongue in place. Here the odds makers will find Cro-Magnon's wicked right hook: speech.

As children we're taught the alphabet with a special emphasis on vowels. In English-language classrooms, we learn to say "A, E, I, O, U (and sometimes Y)." In this simple teaching sing-song is the secret of why we became the dominant predator species on the planet: language.

Here is where Cro-Magnon's wicked right hook of evolution KO'd the Neanderthals and their many advantages: Cro-Magnon's ability to orally communicate complex thoughts and ideas.

Conversely, the Neanderthal's could mostly grunt, but not in the manner of stereotypes. Neanderthal physiology produced a slow-paced and nasalized speech, with a very limited range of sounds. It was high pitched and sharper than that of modern man.

The vastly superior capability our Cro-Magnon ancestors had to quickly express complex thoughts and ideas was how they KO'd the Neanderthals. This advantage led to spoken languages, then written languages, the printing press, and eventually to the Internet. Every time we type an e-mail on a keyboard, each keystroke adds another stone to the buried past of the Neanderthals.

Look at what you're holding in your hands right now, whether a printed or electronic book, what you hold in your hands is irrefutable physical evidence of why we're the king of the hill. For now.

"That's all fine and well," you might be saying to yourself, "but what does this mean to me in terms of my own survival?"

This question is more than fair. This is one you better darned well be ready to go down on the mat with.

It is what we can do with visualization.

Here is one of the Neanderthal's advantages; this is one thing they could do marginally better than our Cro-Magnon ancestors. They could visualize the world about them, in their minds. It is why they were the first to bury their dead.

Granted, our Cro-Magnon ancestors possessed lesser powers of visualization, but their deficiency was amply offset by their ability to express that which they had visualized an evolution-winning talent.

In today's modern consumer world, humans have adapted physically to a new environment. As our television screens grow, so do our derrieres. Everything just tends to broaden, because genetically speaking, we're still devious adventurers, scrounging the Earth. But there's an upside!

Each of us has this package buried in our DNA so to speak: A proven survival talent up in the dusty corners of our genetic attic. All we need do is wipe away the cobwebs and dust and open the box.

What if that box has a padlock on it and there is no key? No worries, its time to get yourself a hammer.

Buying a Hammer Visualization

Hammers are very effective force amplifiers and need to be considered when organizing survival gear. During a catastrophe, a simple hammer can give you either a narrow or a wide range of survival options.

Most of these options you have never considered before. However, with a little bit of visualization, those options can come into sharp focus and in a life-saving way. To illustrate, let's begin by comparing two different kinds of hammers; such as those found in any local hardware store:

- Household curved claw hammer: Handy for mounting pictures on the wall, building tree houses, and other general household uses.
- Construction rip claw hammer: Used by construction workers when framing houses, etc.

Now imagine that you are shopping in a local hardware store for a hammer that will serve both immediate needs and long-term survival goals.

After careful consideration, you've narrowed your search to two hammers from the same trusted manufacturer. One is for the home and the other for construction sites, and we'll use a present-day product comparison list to introduce them.

Product Feature	Household Hammer	Construction Hammer
Claw Type	Curved	Rip
Construction	Two-piece, steel head with an unbalanced wood handle	Single piece, drop forged with a balanced handled.
Hammer Head Face Size	100% of Normal	125% of Normal
Hammer Head Face Type	Smooth Bell	Criss-cross Groove
Typical Weight	13 oz (.36 kg)	20 oz (.56 kg)
Comparative Price	$8.00	$28.00

As modern consumers, we naturally want something that gives us immediate value. In this case, the construction hammer is more likely to lose out to what we're most likely to need today a reliable hammer for general carpentry tasks around the house.

In a current sense, this is logical. House hammers are both lighter and considerably less expense than construction hammers. Also, their curved claw is better suited for pulling those tiny finishing nails out of the wall and such.

So the decision goes to the household hammer because it offers value for jobs we need to perform today. In a future sense, a hammer has no moving parts so it stands to reason it could serve as a useful survival tool as well, for example, to drive tent stakes, build temporary shelter, and so on. Oh, really?

Now, imagine these two hammers in your hands in a brutal post-historic world where jobs are not simple anymore. Now you need a hammer that's versatile for post-catastrophe tasks such as rescue, climbing, preparing campsites, foraging, and defense. Of these, the most likely first need for city dwellers will be rescue.

To visualize this, imagine that you're fixing the backyard fence when a major earthquake strikes. Half of your home collapses, trapping members of your family, and you hear them calling for help. You have in your hands two hammers, and you only need one. Here are the differences:

Rescue Application	Household Hammer	Construction Hammer
Claw Type	The curved claw is virtually useless for prying debris because the extreme bend makes it difficult to drive the claw into small spaces.	You can flip the hammer head around and pound the nearly horizontal drive claw directly into small cracks and crevices with ample prying leverage.
Construction	The two-piece design relies on a wood or fiberglass handle that can shatter.	Because the hammer is a single piece of drop-forged steel, it is designed not to fail.
Hammer Head Face Size	The normal 100% head face size is a reliable force multiplier. Here, the consumer-driven decision is acceptable.	Because the construction hammer head size is 125% of normal, it delivers more force with each blow, for a faster rescue.
Hammer Head Face Type	The smooth bell face design is ideal for finishing nails. For rescues, a smooth face hammer is more likely to glance off.	The criss-cross groove design prevents glance-off and ensures that the full force of each blow is delivered, for a faster rescue.
Typical Weight	The typical weight of 13 oz (.36 kg) is principally in the head of the hammer. This means the unbalanced shock of the impact will pass through the handle into your arm, fatiguing you faster.	The perfectly balanced one-piece design delivers the full force of 20 oz (.56 kg) of directed steel on the broken concrete, walls, and whatever else is between you and your loved ones.

Rescue Application	Household Hammer	Construction Hammer
Comparative Price	Will saving $20 today save a life tomorrow?	Is $28 worth knowing you have a future rescue tool?

With this comparison in mind, let's take a quick look at the other four, post-catastrophic event applications: climbing, building campsites, foraging, and defense.

For climbing, if you're trying to make your way up a steep hill, your household hammer will be dead weight unless you need something to drive stakes. Otherwise, the construction hammer offers a claw that will dig in and strength you can depend on when using it to scale a near-vertical hillside or embankment.

In your mind, see yourself dangling from this hammer. Would you want to dangle from a two-piece hammer with a wood handle that could splinter at any moment or a single-piece forged hammer designed for heavy abuse?

Then there is the campsite pitching tents, building a fire, and so on. Here, the striking force of the criss-cross groove design hammer face can smash up dry branches and lumber for firewood. In addition to driving tent stakes, a rip claw hammer is also handy for digging a fire pit and prying up stones to line the pit.

With a base camp made, you'll need to begin foraging. If you're searching through buildings and cars, a good pry bar is great to have. Short of that, a construction hammer gives you a solid prying tool and a handy tool for smashing through locked doors.

Please keep in mind that you must be careful not to call attention to yourself while you're foraging because if you're successful, you'll have a backpack with goodies. No doubt, there will be others who want to take them from you or worse. So if worse comes to worse...

As a weapon of last resort, a household hammer will more likely encourage attackers than deter them. On the other hand, if all they have are knives and clubs and they see a raised construction hammer in your hand, their reaction will be much the same as that of medieval knights.

In medieval Europe, wealthy nobleman and knights wore expensive body armor. It made them dangerous because it made them difficult to kill. Consequently, those who could not afford body armor had to find some way to gain a tactical advantage.

One notable weapon with a proven tactical advantage over body armor was the German war hammer. Imagine a modern rip-claw construction hammer with a handle twice as long and a smaller head, and you're looking at a medieval German war hammer. The hammer's longer handle provided good reach but still enough weight to deliver a crushing blow to a metal helmet.

If the hammer found a weak spot, it could actually penetrate the helmet and kill the knight. If not, a crushing blow with this hammer could deliver enough energy through the helmet to kill the wearer or, at the least, disorientate or stun him and therefore make it easier to dispatch him.

So, now return to the confrontation visualization and see standing before you a thug with a hunting knife sizing you up. He'll see your raised hammer and the first thing he'll do is look for signs of fear. If he sees them, he'll be emboldened. If not, here is what he'll see next.

A 20-oz hammer looks like a 20-oz hammer big. The hammer head is oversized (125% of normal) and the criss-cross groove design means that if you land a blow, it will not be a glancing blow. Wherever that hammer falls, the thug's body will absorb the full force of a bone-crushing impact.

Then there is the handle itself another problem for the thug to consider. A two-piece household hammer will use a wood or fiberglass handle that is often squared on the leading and trailing edges. That means if the hammer handle and not the head strikes the thug, the force of an impact from a household hammer can be deflected and is therefore less likely to cause him harm. Ergo, the head of the household hammer is the only significant threat.

However, with a one-piece forged hammer, the handle is thinner and shaped like the bow of a sailboat on both the leading and trailing edges. Now the risks are much higher because the oversized face of the hammer plus the greater length of the handle form one continuous threat surface. No matter what part of the head or handle strikes the thug's body, it can focus enough energy to bruise or shatter a bone.

The bottom line here is that regardless of whether you're holding an $8 household hammer or a $28 construction hammer, you've got to put your game face on and do your level best. Remember, a hammer raised can sometimes be as effective as a hammer swung.

Now, let's hit the pause button for a moment.

Stop and think about what you've been doing for the last few pages.

You've probably been exposed to concerns you've never before considered. Now be honest with yourself. From this day forward, will you ever see hammers the same way again? Did you imagine yourself using a hammer to deter an attack? If you said yes to either question, here's the good news you need.

You're already defending yourself. You've used visualization to transit the first stage of a heuristic process. Along the way, you've learned something entirely new about your own survival.

At this moment, this awareness resides in your short-term memory, which means it is freshness-dated. Down the road, you may remember having read about it, but the memory of what you've read will often be hazy.

When you're trying to save your life or that of someone else, hazy just doesn't cut it. What you need is sharp and clear and that means turning this momentary awareness in your short-term memory into a long-term neural network that's ready on a moment's notice.

The easiest way to do this is with physical reinforcement and the sooner the better. Physical reinforcement is an easy way to turn your visualizations into new neural pathways. For example, think of a time in the next 24 hours for a visit to the local hardware store.

Go to the hammer section in the store and hold various hammers in your hand and imagine using them in different post-historic survival situations. One word of caution, please make sure that nobody is standing next to you, in case your kung fu gets the better of you.

The same physical reinforcement principles hold true for war-fighters, SWAT teams, and the like. It is why so much money is spent training them in realistic, life-like training settings. The physical reinforces the mental and the result is a wetware update to the brain that's always switched on.

Remember, this is a heuristic process so the more often you repeat it, the more comfortable it becomes, and you'll get progressively better at doing it as well. Hammers are a good a place to start because they are widely available and inexpensive survival tools. Start doing this today and, as you do, your confidence in yourself and your ability to survive a global cataclysm will be your reward.

You'll also begin to see the world in a fresh new way and you will appreciate it more than ever before. Also, once you've freed your mind of consumer indoctrination with regard to surviving the tribulations, you'll also understand all too well why it will be a horrific taker of lives. Free your mind and your fate will follow.

One way to free your mind is to visualize hammers along with everything you'll need to survive within your own five zones of control. When you accept this challenge, you must resolve through a personal commitment that regardless of how many heuristic trial-and-error mistakes you make, you're going the distance.

Here is even more good news you need.

The minute you make a commitment to building your five zonesof control, you immediately cease to be a victim. You become an Earth Pilgrim, born to survive what comes.

The Five Zones of Control

Whatever is happening elsewhere in the world, it is likely going to happen, with or without you. Even if you could control it, there is no time to make a difference, so why fixate on it?

Rather, get on with the more pressing business of survival and start using visualizations to prepare yourself for that which is within your reach your five zones of control.

Zone	Description	Guiding Principle
1	Personal	Choose your steel wisely.
2	Encampment	Plan for the weakest.
3	Outer Perimeter	Everyone defends it.
4	Observation and Foraging	Move like a ghost.
5	Electronic and Supernormal	Build connections of hope.

The hammer exercise discussed earlier is just one example of the kinds of visualization you need to perform for each of your five zones of control. As you develop visualizations for each of your five zones, keep the following tips in mind:

- Let it come naturally. Do not force the visualization. If it does not come at first, be patient and persistent and it will come. Once visualization begins to occur, it's like learning to ride a bike. You never forget how.

- Prepare to make mistakes. While you have time to make mistakes, make the most of it. If you allow yourself to become frustrated and resigned, you'll retreat into your consumer indoctrination and inevitably fail. Take responsibility for your mistakes and what you learn will help to sharpen your powers of observation and improvisation. Then, better your odds.

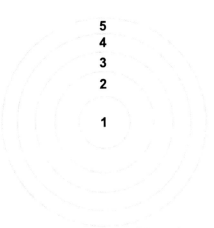

Your Five Zones of Control

5
4
3
2
1

Illustration 55: Zones of Control

- Allow your confidence to grow. If there is anything you want to know about naysayers, go ask Noah. He had more than his fair share of them. If you let them, naysayers will rob you of your confidence, thereby dooming you both. Stand in your knowing, and alone if you must.

- Look for the critical technology. A knife can be as fancy you wish, but the critical technology is always the blade's edge. Is has to cut well, hold an edge, and sharpen quickly. Everything else is secondary. Remember, the critical technology always drives the decision.

The five zones of control are not intended to become a movable spiritual retreat. The zones of control are where you live and survive. Stay within your five zones of control and as for the rest of the world, Max Ehrmann's Desideratasays it best, "...whether or not it is clear to you, no doubt the universe is unfolding as it should."

Zone 1 — Personal

Guiding Principle: Choose Your Steel Wisely

The actual size of your personal zone of control is defined by your weapons, whatever they may be, your skills, and your determination. To help you visualize the diameter of a personal zone, look at it from the standpoint of self-defense.

For example, with a stand-off weapon, such as a pistol or spear, the diameter of your personal zone could be as much as 25 yards (22 m). With a hand-held weapon such as a hammer or knife, the diameter is roughly equivalent to your height. If you are 6 feet (1.8 m) tall, six feet will be the diameter of your personal zone.

Right up front, the most important thing you need to remember about a pistol is that if you need to draw one, you've already screwed up, because you did not handle the threat from a safe distance or, better yet, steer clear of it altogether.

This is because when a threat enters your personal zone, your chances of sustaining a life-threatening injury are dramatically heightened. In Westerns, when the good guys are grazed on the arm, they wrap a sweaty bandana around it and come back shooting. One good scratch is always worth a half dozen bad guys.

However, this won't be true during a tribulation when access to medical supplies and services is heavily taxed and common things like antibiotics are in short supply. Then, a simple Western shoot-em-up good-guy wound could easily become

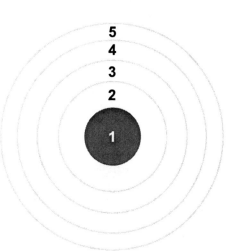

Illustration 56: Zone 1 — Personal

infected in a filthy catastrophic environment. In that reality, once your good guy begins to smell like cheese, you're on your way to an American Civil War-style, field-hospital amputation or death. There will be nothing romantic about it. Nothing at all.

If you want to live, make it a point to deal with your problems at the greatest possible distance. This is why people in the Old West carried "belly guns," pistols tucked in their belts or worn in a holster. They were their last line of defense if they had failed to deal with what was coming at them at rifle range and what concerned them most were animals.

This is why they valued good rifles over pistols and why their less powerful black-powder guns had such large calibers. They were not consumed with worry about gunslingers or renegades; they wanted animal knock-down power in both their rifles and their pistols.

In the coming tribulation, you'll want animal knock-down calibers for the very same reasons. Yet, here is where many women will feel an abiding reluctance one that will put the issue on hold. With this in mind, let me take a moment to speak to you ladies out there.

Philosophically, it is my belief that if every woman in the world could kill a man with two fingers, we'd have a better world warts and all. It's not about which sex is more capable of leading, it is about evening the odds. Ladies, the same holds true with arming yourself for the future. It is about your evening the odds before the die is cast. If you sit in the wings and wait to see how things play out, you'll be in a place where death wears fangs.

In the future, domesticated dogs will become a food source for us, and vice versa. After the small breeds are devoured by large breeds and natural predators, more pressure will come to bear on humans. Weakened by disease, dehydration, and malnutrition, humans will be easy prey, and large feral dogs and natural predators will pursue the advantage.

This is not something that may or may not happen in some distant future, as some may reason. We have proof of it today, and a tragic example illustrates the kinds of future risks to women and children a tribulation can be the violent death of 28-year-old Yovy Suarez Jimenez on May 11, 2006.

Having gone out for a daily jog along a bicycle path that ran along a canal near her home in Davie, Florida, she was sitting on a bridge over the canal when she was attacked. Apparently, she was dangling her feet off the bridge when 9.5-foot (2.9 m) long bull alligator bit into one of her legs and dragged her into the canal.

Later, when searchers found her body, she was still wearing her Nike trainers, sports bra, and biking shorts. But both her arms had been ripped off. When the bull alligator was later found and destroyed, the two arms were found in its belly.

Even though the alligator had dragged her into the water for the kill, there was no indication that she had drowned. The autopsy showed that Yovy had died quickly due to the trauma of blood loss and shock. Because there were no eyewitnesses to the attack, one can only presume how the attack played out.

For the sake of argument, let's assume that this pretty 28-year-old woman was fighting for her life right to the end. In your mind, see here pounding her clenched firsts on the head of this monster with all the strength she could muster. Perhaps this is why the bull alligator went for her arms. He was attracted to the movement.

Illustration 57: Yovy Suarez Jimenez

Ladies, during the tribulation you will see attacks like this, and not once every few years. For a period of time, they will happen every few days, or perhaps daily. Now ask yourself whether your present sensitivities about firearms are worth the possibility that one day you will be attacked by a voracious predator and your only defense will be two clenched fists like Yovy Suarez Jimenez.

Another reason to think about this today is that in a tribulation, predators will eye you and your children as compassionately as we do a take-out bucket of fried chicken a mobile feast where they need only single out those least capable of saving themselves, much like we like to pick low-hanging fruit.

If you are ready to accept the responsibility for arming yourself, then visualize a time when these predators press their attack. You'll be lucky to get off more than one or two shots, which is why you need simple stopping power, for both near and far. There is also another reason.

During tribulation, some people will go insane and pounce indiscriminately on defenseless targets. Like bloodthirsty crack addicts, they'll possess extraordinary strength and be virtually oblivious to pain.

This is why they'll be as hard to stop as an alligator, bear, feral pitt bull, or other predator large enough to take down a human. Ladies, you'll need animal-stopping power to deal with them. If you're having a hard time learning this lesson, you're in good company the US Army.

In 1911, the US Army adopted the Colt .45 ACP pistol after a lesson hard learned and, ladies, this lesson will help you give you added perspective.

Prior to 1911, the standard pistol issued to officers was a smaller .38 caliber revolver, and what brought about the change of calibers was the Philippine American War (1899-1902).

During the war, Philippine guerrilla fighters were poorly equipped but highly motivated. Consequently, some would prepare for an impending attack by first binding themselves tightly around the midsection and then drugging themselves into a painless stupor.

When an American patrol entered their ambush zone, they would spring out of the grasses with fearsome bolo knives in their hands. Their targets of first choice were the officers because the guerillas knew that the .38 caliber pistols the officers carried lacked stopping power. They also knew that a well-aimed shot at a moving target is challenging, especially in the heat of battle.

Consequently, US Army officers had to get a good head or heart shot or risk being hacked to death, and far too many met their end because the .38 caliber pistol proved itself to be a sad failure. That war is remembered as having one of the highest death rates for American fighters of any war ever fought by this nation.

The result was that the Army set about the business of arming officers with a new pistol. They needed one with real stopping power and so the "official" specification called for a pistol that could kill a horse. While that was a good explanation for publicity's sake, the real intention was to stop tightly bound, drug-crazed enemy combatants.

The Colt .45 ACP proved it could do the job and was adopted in 1911; it was carried from WWI until it was replaced in the mid 1980s with the NATO 9mm. Furthermore, the .45 ACP remains the weapon of choice for many firearms experts because of its reliable stopping power, which is considerably greater than a 9mm round. A .45 ACP will stop a charging animal where a 9mm will not, unless you're a very good shot.

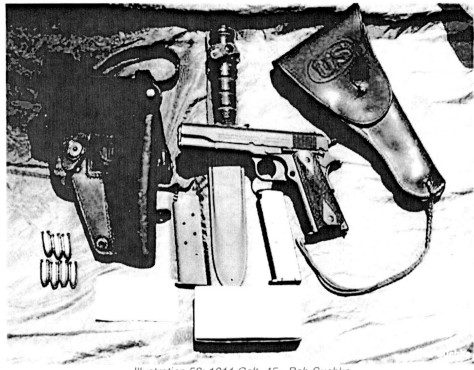

Illustration 58: 1911 Colt .45 - Bob Suchke

Personally speaking, the Model 1911 Colt remains my own personal choice, but that is not to say that it is the right defense weapon for you. This is because your choice needs to reflect your needs and you need to research it seriously.

This means that buying a flashy pistol on sale and tossing it in a drawer with a box of cartridges is a good way to get yourself killed. This is because toss-it-in-the-drawer gun owners are all too often amazed to find that they cannot hit their targets, even at close range. Ergo, a lack of proficiency is like a small caliber cartridge. You better be damned lucky.

Also, forget sexy. A self-defense weapon is not a bragging right. It is not a political expression of how we want to see the world; it is a tool and nothing more. Exotic pistols such as those we see in the movies are cinematically exciting, but finding ample supplies of ammunition and parts for them is less exciting and darned expensive as well.

The point here is that given the time remaining for preparations, the best place to start your search for what works best for you will be with combat-proven technologies. With guns, the critical technology you need to be mindful of is the caliber of the round. It defines the force that meets the target and at what range, so this is where you need to start.

For best results, begin with proven knock-down calibers that are in wide supply. While gun enthusiasts love to endlessly debate calibers, the simple fact remains that a rifle without ammunition is, at best, an awkward club.

With this in mind, here are some popular caliber sizes as reference points for your own examination. From these calibers you can explore your options in a way that makes sense for your survival needs. For men, start with the .45 ACP and 44 Magnum. For women, start with .357 Magnum and .40 S&W. Then experiment from there.

These four calibers offer more stopping power than a 9mm or a .38 caliber round, and they will give you an opportunity to try both revolvers and clip-fed, semi-automatic pistols. Check your local area for an indoor shooting range that offers both classes and gun rentals. Then go learn, and try before you buy. Remember, whatever you choose needs to be something that fits your hand, and you must trust it to be accurate and reliable.

There are also other personal considerations in choosing your steel wisely. There are the obvious things like knives and hammers, but you'll also need needles for mending clothes and wounds, whistles to call for help, fire starters, binoculars, and other items. Collectively, these will become your personal "steel" as an Earth Pilgrim. To illustrate the point, let's start with whistles.

There are two popular types of whistles used today those used in sporting events and those used by the police. The P-shaped sports whistle is louder than the long tubular police whistle, but is it the best choice? Consider this; if you're buried in rubble and caked in dust, a tubular police whistle is easier to blow. However, since both are small and lightweight, instead of choosing one over the other, carry one of each and make both easily accessible, especially for the children!

A similar example is fire starters. Two popular types are small magnesium block starters and steel strikers. You shave magnesium block starters with a knife to make slivers of magnesium kindling, which makes them a very handy tool in rainy weather.

On the other hand, a steel striker will last much longer and works as well as a magnesium block starter, with a little extra practice.

Like whistles, fire starters are small and light, so the same rule applies. Carry both types. Use the longer lasting steel striker for most situations and save the magnesium starter for more difficult times.

In terms of managing your own personal zone, the bottom line is this. Water comes and goes. Food comes and goes. Shelter comes and goes. People come and go. Your steel will be your steel and you will carry it always. Therefore, when you go shopping for your personal steel, leave home without the bucket lists and choose wisely.

Zone 2 — Encampment

Guiding Principle: Plan for the Weakest

The term "encampment," in a military sense, describes a place with temporary accommodations consisting of huts or tents. However, the same term applies to nomads, and in the case of Earth Pilgrims, you must plan for the weakest.

During the tribulation, life on the surface will be subject to events and the environment. Consequently, until things settle down, Earth Pilgrims should expect to live from one temporary encampment to the next, much like the American Plains Indians before they were corralled into hopeless reservations.

Like the Plains Indians, Earth Pilgrims will organize together into families, clans, bands, and tribes, and roughing it in the bush like a squad of soldiers will not work.

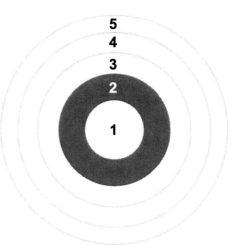

Illustration 59: Zone 2 — Encampment

Earth Pilgrims will gather together in a mix of men, women, and children of various ages and states of health, so they must have different ways to plan encampments. Consequently, when Earth Pilgrims select a possible encampment site, their concerns will span a broad range of needs. This is why it is imperative to visualize how you will empower the weakest among you, because each one who is able to contribute will have at least one useful strength.

In that future time, what we perceive as useful strengths will change as well. Consider a grandmother who is very skilled at making herbal remedies from local plants and grasses. This is indeed a valuable talent. This grandmother is not weak in terms of her value to the group; rather, she is a treasured asset. However, if worse comes to worse, that grandmother must be ready and capable of defending the encampment from predators or marauding thieves.

Twelve-gauge buckshot is the ammunition a grandmother needs, but the type of shotgun best suited for her is not the type we see mounted in police cruisers—pump shotguns with a short, combat barrel. Handling a shotgun like this requires familiarity and physical strength, especially if there is only enough time to get off one or two shots.

On one hand, you could spend an afternoon and considerable amounts of ammunition teaching grandmother to shoot a 12-gauge pump shotgun. Or, you could plan for the weakest. While strong men and women can carry 12-gauge pump shotguns, what grandmother needs to defend the encampment is a good old-fashioned sawed-off, double-barrel 12-gauge shotgun. A simple scattergun in the hands of dear sweet old grandma will send a unique message to the bad guy.

Look at it from the standpoint of an attacker. A frail grandmother holding a single-barrel, pump shotgun means that she'll probably get off the first shot assuming she has chambered a round and assuming she does not choke while trying to reload. If she looks vulnerable and scared, that opens new possibilities.

On the other hand, a double-barrel 12-gauge scattergun sends a different message to someone assaulting the encampment: "Yo, bucko, I'm such a simple scattergun to use that any sweet old grandmother can use my two rounds to cut you in half, along with a few of your friends. Do you feel lucky?" If the marauder also sees a corresponding "do yah feel lucky?" in dear sweet grandma's eyes, maybe he'll be inclined to move on to easier pickings.

Grandmothers will have no trouble putting on a game face; you just plan for the weakest and put the right tools in their hands. The same holds true with everything else in the encampment, whether it be cooking utensils, clothing, or other supplies. Everything must be durable and simple enough for everyone to operate, beginning with the weakest among you.

Zone 3 — Outer Perimeter

Guiding Principle: Everyone Defends It

Whether you are in a small temporary encampment, huddled alongside a road, or in a railroad tunnel, you must continually keep a vigilant watch on the outer perimeter of your encampment. No matter what else you do, everyone needs to continually scan the outer perimeter.

If attackers see that everyone is preoccupied and distracted, they'll be encouraged to attack. Therefore, if you're vigilant, the risk to the attacker is noticeably high, even if all you have on hand are crudely fashioned wood spears, clubs, and the like.

Remember the whistles? Everyone should wear one around the neck to alert others of an attack. Even if you are unarmed, such vigilance can serve as a deterrent to those spying on you from a defoliated tree line, crumbled building, or other form of concealment beyond your outer perimeter.

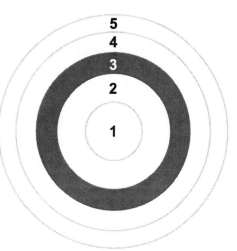

Illustration 60: Zone 3 — Outer Perimeter

If you plan to arm yourself, which you should, a good rule of thumb is to define a 100-yard (91 m) perimeter with clear ground between the outer perimeter and encampment. This will force marauders and attacking animals into the open.

If you intend to arm yourself for this contingency, pistols are virtually useless at 100 yards (91 m) and the latest and greatest and most expensive weapons will likely be too complex for a tribulation survival situation. What you want is simple firepower that is durable, and you want plenty of it.

For the same money as a popular bolt-action hunting rifle, you can purchase half a dozen surplus bolt-action military rifles in serviceable condition from a local sporting goods store or gun dealer. For the price of a modern semi-automatic or automatic rifle, you can easily buy twice that many surplus bolt-action military rifles—and have lots of money left over for ammunition and cleaning supplies.

Yet, isn't an automatic or semi-automatic rifle really the way to go? If you think you can throw down on well-paid and well-armed government soldiers of fortune, you're playing a fool's game. They'll be more mobile than you are and with the first shot, they'll know you're around. With the second shot, they'll know where to look, and they'll cover ground faster than you can retreat. Just don't go there!

The point here is that moving your encampment to avoid well-armed government soldiers of fortune is the smart thing. They cannot be everywhere and in force and eventually they'll lose their power. In the meantime, drawing their blood will only draw more of them to you. If they strike before you can move, the more long-range, knock-down-caliber barrels you have pointed outward from the encampment, the more formidable a defense you present.

Let it not come to that, but if it does, proven 20th century bolt-action military rifles are superb for defending a 100-yard (91 m) perimeter, plus they are qualitatively inexpensive and still widely available. The most popular types, along with plenty of ammunition, can be found in local sporting goods stores.

Therefore, in planning outer perimeter defense needs, the following three popular surplus bolt-action calibers offer a good place to begin your own analysis:

- .303, British Lee-Enfield
- 7.62x54R, Russian Mosin–Nagant
- 7.92x57mm (aka 8mm) German Mauser

All three calibers are generally referred to as bolt-action, 30 caliber rifles and are ideally suited to a cataclysm survival requirement. While the American surplus rifles like the .30-06 Springfield are excellent weapons, they are considerably more expensive. Likewise, as events become chaotic, governments will begin seizing semi-automatic assault rifles such as the AK-47 and AR-15.

So, as you select your tribulation rifles, visualize the following benefits in your mind:

- Proven reliability. The more complex a weapon is, the more it requires modern gunsmiths and modern tools. During the tribulation, the air will be filled with soot, volcanic ash, and dust—a brutal environment not much different from the muddy WWI trench.
- Penetration and stopping power. All bolt-action, military .30 caliber-style rounds are capable of penetrating light armor and delivering knock-down power, and at

further distances than lighter weapons like smaller caliber assault rifles used by the military today, such as the M-16.

- Excellent steel sights. Rifles with optical scopes are accurate but temperamental, which is why hunters always have to "sight in" their rifles before hunting season. During a tribulation, keeping a scoped rife sighted will require precious ammunition and time. Old military rifles come with superb steel sights that are designed for use as far out as 800 yards (731 m) in some cases. Once they've been calibrated at the factory, continual sighting in, as with scoped rifles, is unnecessary.

- Simple to operate and clean. These rifles are simple to operate and clean. Also, most types come equipped with a pull-through cleaning rod, a very handy thing to have.

- Parts interchangeability. When you buy several inexpensive military rifles of the same variant in good condition, the parts are usually interchangeable. Note, however, that this is not the case between variants of the same caliber. For example, the popular K-98 8mm German Mauser and the M-48 8mm Yugoslavian Mauser look similar and use the same cartridge, but the parts are not interchangeable.

- Personal range defense. You can purchase bayonets with these old rifles, and you should. Modern military rifles use shorter bayonets which are good for killing people, but these older riles use bayonets up to twice as long, and they can be used to skewer an attacking animal.

When comparing these three calibers, you'll see that the availability of ammunition differs. While there is a sizable supply of cheap war surplus ammunition for the 7.62x54R, Russian Mosin–Nagant, and 7.92x57mm (aka 8mm) Mauser, no surplus ammunition is presently available in the United States for the .303, British Lee-Enfield, since the Clinton administration blocked the import of the caliber.

The reason why the Clinton administration banned surplus .303 ammunition is that the Lee-Enfield rifle is easy enough for any grandmother or 12-year-old child to operate. Conversely, rifles that shoot the 7.62x54R and 7.92x57mm calibers are intended for the physical strength of a relatively healthy man.

Another reason why the importation of surplus .303 ammunition was blocked is something called the cyclic rate of fire. It is why the Lee-Enfield is considered to be one of the most deadly bolt-action military rifles in the world.

The Lee-Enfield's cyclic rate of fire (how many rounds you can load and fire per minute) is roughly twice that of the other two calibers. At the outset of WWI, the Germans assumed that these rifles were comparable to their own Mausers. They learned otherwise, the hard way.

Another important advantage of the Lee-Enfield is that it uses a detachable 10-round magazine. The other two caliber types store five rounds in their stripper clip-fed, internal magazines, which can be more difficult to reload than a Lee-Enfield. Ergo, in a firefight, a mad-dog-mean grandmother with a fully loaded Lee-Enfield and a clear line of site can be one damned deadly old lady.

However, although inexpensive surplus ammunition is no longer available for the Lee-Enfield in America, it is not a deciding issue here when it comes to surviving a tribulation. This is because surplus ammunition tends to be corrosive. This means you have to clean your rifle before the residue pits the bore of your rifle, thereby permanently reducing its accuracy.

Also, surplus ammunition is not as accurate as modern fresh loads. The difference at 100 yards (91 m) is that surplus ammunition is not effective for head shots, whereas modern fresh loads can be. Nonetheless, they both hit with devastating power.

Therefore, surplus ammunition is a plus for target practice prior to the tribulation, but during that time, non-corrosive, fresh-load ammunition is what you want on hand, and it is widely available for all three calibers at approximately the same cost.

The bottom line with your outer perimeter is that not only does everyone need to defend it, you need to visualize yourself giving everyone the right tools to do the job, and that includes grandmothers, pregnant women, and 12-year-olds alike. Survival is everyone's fight.

Zone 4 — Observation and Foraging

Guiding Principle: Move Like a Ghost

One skill you need to begin developing today is that of observation. As you drive to and from work, the store, the day-care center, and elsewhere, turn off the radio and see the world around you in terms of survival opportunities and perils. This is because when an unforeseen cataclysm strikes, you'll likely be traveling one of these routes.

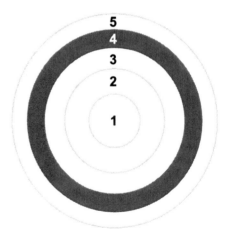

Zone 4 — Observation and Foraging

Illustration 61: Zone 4 — Observation

Look for road junctions that can bottleneck traffic in several directions, and make it a point to not only explore alternate routes, but to travel them when possible. When a water main breaks or road crews are tearing up the street to lay a new gas main, take note of how these everyday events can gridlock traffic for blocks and sometimes further.

During a major earthquake, gridlocks like these will be one of many obstacles separating you from your home and loved ones. This would not be a good time to become paralyzed with fear because you've been caught completely unawares and to have no plan of action but to say, "OMG, what do I do next?"

Also look for anything that can serve as a natural radiation shelter. In Planet X Forecast and 2012 Survival Guide, solar radiation was presented as the greatest threat, and it remains so.

During a solar storm such as those predicted by the Avebury 2008 formation, if you're caught out in the open in sunlit portions of the Earth, you'll have a few minutes at most to get to shelter before you become the proverbial "poodle in a microwave oven." The radiation may not kill you immediately, but over time it will do you in.

Also note that solar storms usually come with a one-two punch. First, the radiation reaches us at the speed of light; it reaches Earth in 8.5 seconds, long before you can hear the sirens beginning to wail.

In fact, before you hear the sirens, you'll see the modern digital technologies used in cars, radios, and other devices stop working. The first thing the radiation will do when it strikes the Earth is to cook these devices in their own juices in a fraction of a second.

This is why having an MP3 player on hand is useful in the tribulation.

Any cheap MP3 player will be your canary in the coal mine. When it suddenly stops working, you don't tap on it or twist the earphone jack, YOU GO TO GROUND. You get as much rock, dirt, sand, concrete, and steel over your head as possible. So, if you can find cheap MP3 players with removable batteries, stock up. At the least, do not throw away the old ones lying around the house.

While commuting to and from work, memorize the location of all the overpasses, underground parking garages, and culverts. These will be the places to sit out the radiation wave from a solar storm. After that, you have a matter of hours to get your loved ones to safety; the gases of a coronal mass ejection (CME) follow on the heels of the radiation for a complete one-two punch.

Start finding and remembering these places today; do it quietly and move like a ghost. If you tell everyone what you're doing, some day when you need these places, your own gossip may result in there not being enough room for you.

Remember, in a survival setting, a chatty Easter egg hunt approach will be your undoing. Be aware, move like ghost, and keep notes if possible.

Zone 5 — Electronic and Supernormal

Guiding Principle: Build Connections of Hope

In the present day, our lives are filled with connections of many kinds, both human and electronic. After the solar storms predicated by the Avebury 2008 formation come to pass, what is unimaginable for us today will become the retrograde reality of the future a future reminiscent of life in the early half of the 20th century.

Nor do we need the Avebury 2008 formation to tell us this is coming. Mainstream scientists are already sounding the alarm. They are warning us that the solar storms during the coming 2012 solar maximum will be powerful enough to wipe out much of our global communication networks and power grids, thereby paralyzing our modern societies.

Therefore, the relevant question here is not how we survive a paralyzed global communication grid, but rather how we use the communication abilities that still work. In time, when things are settling down in the aftermath, surviving ham radio operators will dig out from their attics antiquated, vacuum-tube radios and other outdated analog technologies that can survive the radiation of solar storms.

Zone 5 — Electronic and Supernormal

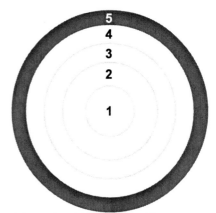

Illustration 62: Zone 5 — Electronic+

Between those two points in time, we will need to open our minds to all conceivable forms of communication, whether electronic or supernormal. If it helps, no matter where it comes from, use it. Do not let modern expectations and assumptions cheat you of talents and technologies that you would not normally consider.

A good example is someone who has been blind from birth. Let's assume you're organizing a survival community in advance of the tribulation. You might wonder whether it is wise to bring someone who has been blind from birth into your community. The question is legitimate given that the blind present a range of specialized planning issues, such as tasking someone to look after the blind and determining what kind of drain this will be on the group as a whole.

Personally speaking, if a disabled person, blind from birth, came to me and asked to join my survival group, I would say, "You betcha. We need someone just like you." Does that sounds odd? If so, let's run it through a few visualizations.

During the day, if only one MP3 player / radiation storm detector is available, it goes to the blind person, because his or her hearing is more sensitive. A lower volume setting on the MP3 player means longer battery life, plus the blind person is not distracted by sight.

The instant the player dies, a blind person is more likely than a sighted person to notice, and that person will be more attuned to hearing the difference between an electronic glitch and a digital device that has just been cooked in its own juices.

If there is enough food to keep a watchdog for the nights, maintaining vigilance is easily. When the dog barks, everyone wakes up. However, if you do not have a watchdog, a blind person is your human equivalent. Keep in mind that during the tribulation there will be little or no moonlight for sighted people to rely on. This will be the result of smoke, ash, and dust in the atmosphere, so expect the nights to be as dark as a psychopath's soul and filled with danger.

In this setting, a blind person can hear things a sighted person will miss, such as a thief sneaking up on the encampment. And if this blind person has a seeing-eye dog, you've got a catastrophic environment, nighttime threat surveillance system that money just can't buy. Consequently, you will love the blind person and the seeing-eye dog for what they can do, and they will love you for the chance to contribute.

This is just one way to build connections of hope for today and tomorrow. Another way is to listen for the faint whispers of distant friends who dearly love us and are ready to do what they can to help us succeed.

11

We Have Friends

We do not believe we have friends because we're taught to think that we don't, no differently than the way our indoctrination programs our consumption behaviors. The problem with the statement is not that most folks will agree with it. They will. The problem is that the problem itself subtly eludes us.

To illustrate the point, let's examine the following proposition:

```
The only way to see the big picture is to walk
backward until it comes into focus.
```

Does the proposition mean that to look backward is a way to see the greater truth? Does looking backward mean that whatever happened in the past is as important as today? Is there is a third meaning?

There is.

Both initial questions "look backward," but the actual proposition reads "walk backward." The reason for the difference is in the underlying assumptions. Therefore, the third meaning of this proposition is a result of how we are indoctrinated.

We're taught to respond to code words which trigger impulse assumptions that instantly remove us from the actual substance of the communication itself.

This is not done to make us better people. It is done to make us useful to others. Does this mean that assumptions are awful and should be done away with? Yes, but only those used against us.

Therefore, whenever an assumption interferes with your ability to evaluate a message, test the assumption. If you cannot explain to yourself how you arrived at the assumption, it means you probably never put it there in the first place. Move it to the possible-indoctrination column and stay focused.

However, when we can create assumptions as tools of awareness in the pursuit of knowledge, they can be useful to us. Much like the four corners of a jigsaw puzzle, they give us good starting points for visualization, and the next proposition.

- **New proposition:** We are not alone; we have never been alone; we will never be alone.
- **Assumption:** How can we see the night sky and assume that we are alone in the galaxy?
- **Visualization:** A very starry night.

A few years before my first out-of-body experience, I witnessed a UFO on a very starry night in the Prescott National Forest in Arizona, as did two good friends.

It was early summer in 1972 and the air was crisp and clean. With the campsite in order, we all lay down on the ground, head-to-head, to enjoy a brilliant meteor shower.

What a stunning sky it was, too. The meteors were just a bonus because the stars were so thick in places, you could imagine yourself jumping from one to the next.

Against this brilliant background, meteors burned through a third of the sky and the three of us pointed to several of them, calling them out as they fell off into the far horizon. Then one of the meteors did something quite spectacular.

We all caught sight of it at the same time, and we were pointing at it and talking excitedly when the impossible happened. Just near the edge of the far horizon, the meteor instantly reversed course and flew toward us to a distance of approximately 1,000 feet (304 m) directly over our heads. We drew a collective breath. This object's next move was equally impressive. Without slowing, it instantly turned 90 degrees away from us and flew off toward the horizon. The turn was as precise as a carpenter's angle.

Whatever this object was, it had used a very starry night with a meteor shower to mask its entry, reversed course toward its next way point, and then made a right-angle turn and zoomed off. In other words, it employed intelligent control.

This was an amazing and memorable event to savor and, in hindsight, I dearly wish my friends and I had seen it that way at the time.

Rather, we hesitatingly asked each other, "Did you see that?"

Of course, there was a quick round of "oh yah, I saw it," then a long pause.

Finally I asked, *"Do we really want to report this?"*

"No way."

"No."

From that moment on, the three of us never spoke of it again. Oh what a sad waste of a brilliant moment that was—a bitter reward for my regrettable question.

We could have been up till dawn having what could have been one of the most fascinating and memorable experiences of our lives, but for indoctrination. Oddly enough, however, had we not reacted in a way useful to someone else, we might have just let the moment slip through our fingers through neglect.

Yet, the memory of shame we all felt when we each thought about reporting the sighting is what has stuck with me for all these years. We had responded in a way that was useful to someone else.

With this, we return to the first proposition:

```
The only way to see the big picture is to walk
backward until it comes into focus.
```

Experience has taught me that when we do this, those thousands of tiny, colored LED lights dotted along large black rectangular lighting panels will eventually blur into a brightly lit end zone scoreboard that says, "The Big Picture."

The Big Picture

Assume for the moment that we're standing together, mid-field, at the 50-yard line. Before us is the big picture scoreboard in the end zone. After all the years and the miles, for what it is worth, here is what I see. Humankind is doing what all species must do: adapt, evolve, or perish.

In Chapter 10 – We Can Do This, we saw how the first-generation Deep Blue supercomputer could win a game of chess against World Chess Champion Garry Kasparov in 1996 but lose the match. Then in 1997, the second-generation Deep Blue supercomputer did win the match.

Had Kasparov been able to prevent the development of the second-generation Deep Blue computer, he would never have been defeated. Assuming Kasparov could halt the evolution of Deep Blue, would he have done so?

This is the predicament humankind now faces. You could say that we're nearing the end of our own Deep Blue first-generation design and that now we're at the cusp of evolution. If we can cross into the second generation, the old masters of the game will rule no more.

However, there is a distinction. As a species, we're not pitted against a single intellect, but rather powerful interests who are doing everything imaginable to prevent us from achieving the next generation of our evolutionary design.

This is why we do not know and will probably never know all there is to know about how things really are. Whew!

It is as though we're all ants in a massive ant farm made up of one-way glass. Others see us as clear as day but all we see is the blank side of the glass. When that glass shatters, will we be ready to see the world around us as it is?

This is the crux of the issue. What are we truly willing to see?

Exploring that notion begins in one place, within ourselves. It is not about who "they" are. "They" will become known in time and then we'll see what must be done. In the meantime, it is about "we" as a species. Who are we and what purpose do we serve in a universe where all things serve a purpose?

We the People

The first three words of the Preamble to the US Constitution are "We the People." These words, penned in 1787, tell us we need to know about extraterrestrials today, because these words cut through layers of disinformation and deceit like something out of a James Bond 007 movie.

Throughout the years, my research and supernormal experiences have all blended into a panorama of destiny and we humans are at the center of it. Or, more specifically, our evolution is at the center of it.

Granted, the arguments of how we came to be in this world vary.

On one hand, we're made of clay in the theological sense. Scientifically, we're a natural progression of evolution, save for that nagging matter of a missing link. Our descendents were so genetically different from Neanderthals that they could not bred with them, though no doubt they tried. Is this because we were engineered to be different?

This begs the question, what if extraterrestrials were to land and introduce themselves as our makers? Likewise, how would we respond to the knowledge that we're a race engineered to be useful to their needs? Isn't this our secret fear? That we'll be enslaved, once again?

Here is the truth of it. All sentient life in the universe is made up of star stuff and we are all bound by the most elemental form of life in the universe—spores. All of it comes from the Creator and nobody owns the patents.

No matter how we came to be, humankind is here and we're working the claim. As responsible and peaceful stewards, it's our world to keep. As violent and irresponsible despoilers, it is ours to lose.

History is just a prelude to this moment and we, as sentient beings, now have the ability to stand up and say, "We the people of Earth shall come together and write our own history, from this day forward." A foolish notion? Certainly not, but the circumstances are nonetheless fraught with risk.

This is the clarion call from friends afar, be they spirit entities or extraterrestrial entities. Some you've already met in the course of this book. Through crop circles and supernormal experiences, our friends from afar send us the same message time and again:

This evolution is ours to lose and, if we do we shall be enslaved for countless generations to come. Then, only after our world has been so brutally raped of its natural resources that it is of no interest to anyone will we be given our freedom once again, for whatever meager time remains.

In contrast, those who would profit from a de-evolution or retrograde of humankind's natural destiny have their repeating messages, as well. They broadcast that we have no friends and that all the gentle warnings we're receiving from friends afar are but the twisted imaginings of desperate or naïve minds. They say we are all alone, vulnerable and wholly dependent on those who rule us.

Grappling with this dichotomy is difficult and disheartening and I understand it well, for it dogged me for several years. Perhaps it was a carryover of my deep regret for having succumbed to indoctrination in responding to what I and my friends had seen on that amazing starry night in Arizona.

The warning does raise the possibility of a dark dilemma, that one day we will see our evolutionary future through an impenetrable glass ceiling. Like ants in an ant farm, we would be free to come and go within the confines of an artificially created world. Not one created for our benefit, but for that of others.

This thought dogged me until the day I finally found a moment of clarity in We the People, and the odd thing is, it took a Pyrrhic victory to do that.

A Pyrrhic Victory

Earlier I described what happened after I began publishing articles and videos about the 2008 South Pole telescope disclosure videos on YouTube. Disinformationists broadsided me with a rather ugly and substantial campaign. But like they say in the movies, "It's the life we chose." You just deal with it and keep an eye on them.

One of their favorite tactics was the use of slanderous videos on YouTube. When they popped up, I would file a *Digital Millennium Copyright Act (DMCA)* complaint with YouTube and, in each instance, YouTube removed the attack videos and sometimes terminated the account as well.

Up to this point, the degree of sophistication in the YouTube attacks was amateurish enough for them to be considered teenaged pranks. However, after my fifth successful DMCA complaint in a row, my attackers did something rather impressive!

They resurrected an unused account that had been dormant for more than a year and, late on a Sunday night, uploaded 244 videos to it in just over one hour. Time stamps do not lie and the average length of these videos was three minutes. Of the 244, 3 were attack videos against me buried deep in the middle of the pack. The other 241were snatches of video taken from all over the Web. Frankly, I was impressed.

OK, so it was DMCA filing time again time to spin up Johnny Cash singing "One Piece at a Time" and get after it. This always started with a comprehensive series of screen captures for the archive. After that, I prepared the complaint and before hitting the trigger, I went to double-check the URLs. When I did, I found YouTube had already terminated the account for a terms of service violation, and that was the last I ever saw of that particular disinformation attack strategy.

So what happened? Perhaps the best way to answer that is, "Have you ever read a Western novel by Louis L'Amour?"

What's great about L'Amour's books is that the good guys always wear white hats and the bad guys always wear black hats, and what I see out there reminds me of his books. The difference is the white- and black-hat characters in this perpetual challenge of give and take are all anonymous.

Yet... beating the black hats is a Pyrrhic victory, because they still achieved their goal. They still distracted me from more important pursuits.

This would not do, yet where was I to go from here? I needed a single point of truth to help give me clarity. This need was on my mind when a few days later the last channeling session with Betty occurred.

The Wisdom of Serapis Bey

In Chapter 7 - The Meek, I introduced a psychic from Canada I've called "Betty" for the purposes of this book. In her second-to-last session, she had a near-burnout experience and still insisted on doing the last session.

For me, this last session with Betty has always been the most memorable of all the channeling sessions we conducted over a period of 18 months, because it transformed my understanding of the big picture.

Each psychic has his or her own style and preferences, and Betty is especially gifted in that she is a receptive conduit for the other side. In this last session, the entity who came through identified himself as Serapis Bey.

The session flowed very well and, after the introductions and a few brief questions, I told him about the disinformation attacks and that I was looking for knowledge. Serapis Bey's answer was truly wise and kind.

He explained it this way: The history of humanity is like a book of many chapters and the one we are in now belongs to the powerful. They own it and there is absolutely nothing we can do to change that. Therefore, spend the least amount of energy possible in dealing with it.

Serapis Bey then pointed out that this particular chapter is also drawing to a close, as do all chapters.

The next is yet to be written and the powerful know it is not theirs. It is ours. However, they can trick us into sacrificing it. Only then can they win. To prevail, we must leave the past behind and begin thinking about that next chapter. Our chapter. The one of the meek who inherit the Earth.

After a few more questions, we said our goodbyes and I thanked Betty for this last channeling session with Serapis Bey. For me, it was a vital turning point that renewed my hope for the future and gave me what I'd been looking for: clarity.

As a direct result, I changed my approach to dealing with the disinformationists. Instead of fighting them, I would now follow them and this subtle transformation has richly empowered my research abilities. Nothing is more sublime than reluctant dots connecting together in your mind.

We're All Going Somewhere

After 18 months, numerous psychics, and hundreds of hours of recorded sessions, it all boiled down to one essential thought. All the entities we contacted, no matter the type, had something in common with us. Just like them, we're all going somewhere, but on a cosmic scale. This brings me back once again to the first proposition:

The only way to see the big picture is to walk backward until it comes into focus.

What is different about us and other entities is that we're all in different states of existence, somewhere in the grander scope of things. Occasionally we can bridge these different states and communicate. In the coming tribulation, those who survive The Great Winnowing will become significantly adept at this.

So how do we recognize these friends from afar today?

First, you must trust your instincts. It makes no difference whether they come to you in a dream or land on a lonely country road just beyond your headlights or come in any other supernormal form of contact for that matter. Trust yourself to know them and, like an owl, listen with your eyes. Then be willing to stand alone in your knowing if you must.

This brings us to the need for a rather critical question. With this coming evolution that is ours to lose, we already know what failure means to us and to our enemies. But what does our success mean to our friends from afar?

About and Why

A term often used by the guides to describe our psychics was "vehicle." It turned up time and again and, given that our protocol favors trends over specifics, it really stood out.

To help understand it, visualize yourself as a vehicle for a spirit entity. For starters, whatever makes you tick was here before you and will be here after you. In this context, the term is correctly applied. We are indeed vehicles for our spirits and here is where our evolution proceeds at two levels: sociological and physiological.

In essence, our Homo Sapiens ancestors represented a first-generation modern man, like the first-generation Deep Blue supercomputer. Since then, we've evolved sociologically, which is essentially a lot of wetware programming. Now we're at a point where we need to go to the next generation and that will require a big sociological and physiological leap forward if we're to win the match. But isn't that a bit painful?

Consider this. Both the hardware and the software of the second-generation Deep Blue supercomputer were heavily upgraded as a result of losing the first match. Without that first loss, a subsequent winning design could not have been possible. As Nietzsche said, "What doesn't kill you makes you stronger." What Nietzsche didn't say is that the people who usually tell you this are the same ones who are trying to kill you.

Nonetheless, assuming that IBM had only upgraded the hardware for the second-generation Deep Blue supercomputer, could it still have beaten a man? No, because more sophisticated software is what won, with the help of more powerful hardware needed to run it. Conversely, without this more powerful hardware, creating the software needed to win would have been a pointless endeavor.

This brings us to the calamities we face. They will reduce our numbers dramatically, but in turn they will transform us at every level, including the physical. After that happens, our species will become an improved vehicle and our future physical bodies will become capable of hosting more evolved spirits and much more.

This is our destiny as a sentient species to fulfill. It is just as straightforward as building a second-generation Deep Blue with more processing power, memory, and so forth. The Creator of all there is designed us to evolve. Even if we are a bio-engineered species we have the right to be here, which is why we face great danger.

Great Danger

What is the consequence of leaving our evolutionary destiny to be stolen from us? The dreadful possibility that humanity becomes not just a slave species, but minions of evil. A Rome-like militaristic stellar species subordinated to the manipulations of a dark few. A curse to peaceful species elsewhere in the galaxy.

With such extreme possibilities in the offing, a noble evolution or an ignoble perversion, stripping away the romantic aspects of this rhetoric is vital. It is time to see things as they are.

We as Homo Sapiens are already in the process of becoming something new on a physical level. The successful result of that process is that we will embrace peaceful and loving races as a free stellar species. Or, we will become a pathetic waste of destiny.

This is why the guides are always there to help us. In this testing crucible of tribulation, we as a new and peaceful stellar species must seek to align ourselves with like-minded species in a spirit of peaceful co-existence.

And, yes, there are the dark nasties and yet, oddly enough, herein is the sliver lining inside this dark cloud called Planet X.

...say what?

In Chaos Is Opportunity

Think back to the time of the Exodus and The Kolbrin Bible accounts of what happened during the previous flyby of Planet X. It caused the Ten Plagues of Exodus and threw the mighty Egyptian empire into chaos. Within that chaos, Moses seized the opportunity to secure the freedom of Jewish slaves from Egyptian bondage.

Had it not been for the Ten Plagues of Exodus, any efforts to free the Jews easily could have been crushed by an empire in its prime, and in time the Jews would either have been assimilated or disappeared as a people. The point here has nothing to do with the rights of the Jewish people to exist, the punishment of a hard-hearted pharaoh by God, or any other judgmental and subjective reason. It is about control.

In the chaos caused by the flyby, the ability of the pharaoh to exercise his control was heavily compromised. In such conditions, opportunities arise, and when an opportunity arose for Moses, he moved on it. The lesson for us is to follow the wisdom of Moses.

In the midst of all of our suffering and dying during the coming flyby, it will only take a few to understand how the control of elites will become heavily compromised, as it was for the pharaoh of Exodus.

However, there is a noticeable difference of scale. It is one thing to talk about freeing a people and quite another to free a species. Still, this is a window of opportunity that is long in coming and brief in duration.

As the elites and their minions seek the shelter of their manmade survival arks, they will leave behind a surface government to maintain control. They'll still pull the strings like absentee landlords, but they nonetheless will be physically absent.

For them, the role of the surface government is to provide a rear guard for them and to maintain control of the above-ground situation. Expect them to be somewhat successful in this as they have had quite some time to prepare. There will be chaos and they will manage it heavy-handedly, as usual.

However, The Great Winnowing within the cusp will be the one factor that injects so much chaos into their plans and organizations that they will become as vulnerable as the pharaoh of Exodus, if not more so. This is because the last flyby, unlike the one that caused the Great Deluge (Noah's flood), did not cause a pole shift.

Therefore, those who survive the cusp will be very different but there will be the residual vulnerability of indoctrination enough at least to confuse survivors into poor decisions. This is the central thrust of the elite strategy for regaining full control once they return to the surface.

For all their daunting force, the elites will lose their control during the chaos of a pole shift, and they know the odds are not in their favor. Therefore, their plan is to trick above-ground survivors to slip back into old habits, as they always have in the past, but this time they'll be vulnerable.

Defeating them once and for all will only take a scattered handful of those who understand what is at stake and who'll teach other to stand up and say, "Never again."

Is this a role ordained by the Creator for a select few? No. This is what you came to this world to do and this is your covenant with the Creator, whether you choose to accept it or not. It's the whole point of what is coming. The choices we make here are the biggest ones of all.

Be in It for the Species

At some point in your own contemplation, you must decide how you are going to meet the challenges of this tribulation On one hand, you can face it with hackneyed elitist rhetoric such as: greed is good, save your own ass, dog-eat-dog, survival of the fittest, life is unfair, ….yadda-yadda.

On the other, the day comes when you hold out your hands, look up at the sky, and say:

"Today, I take personal responsibility for the survival and freedom of my species and I will do it alone if need be, and without any expectation of reward or recognition. This is my covenant with the Creator to which I bind my own life."

On that day, you will begin to walk in the footsteps of Moses and you will not walk alone. Though most cannot accept this burden of responsibility, they will follow the few who do and they will give their full measure to the effort.

For the very few men and women who assume the responsibility of this awareness, the hopes and dreams of countless generations depend on you, your love, and your courage. By your hand, your thoughts, and your deeds, you help a promising young sentient species to leave behind its violent predilections and, like a butterfly emerging from the chrysalis, evolve to a more graceful existence—

one shaped by harmony and peaceful co-existence. We cannot make it a peaceful universe, but we can exist peacefully within it.

Granted, there are sacrifices to be made and as you lie down for the last time, the touch of a loved one may not be your fortune; rather, it may be a lonely and austere setting.

Even so, you will know that you kept your covenant with the Creator. For that, your legacy shall be a life that is admired for eternity.

When your heart is true to this purpose, you will know it by those who reveal themselves to you. You will sense them and they will sense you. Like will seek like, and in the first moment you will know. This is when the special children will begin to find you.

The Special Children

To us, the coming tribulation will be a new and devastating event for humankind to experience, made worse by the fact that we'll be the last to know. This does not mean that we are unprepared, though, because Mother Nature does like to cover her bets.

Think back to the Redwood trees and their 66 chromosomes. Within all that genetic density is millions of years of adaptation. Consequently, when confronted with an extinction event, this ancient species can draw from its own deep genetic survival well. But what about us humans? As a species, do we also have a deep genetic well to draw from?

It is widely reported that the human genome has been fully mapped, even though more than 98 percent of all DNA is uncoded. Consequently, this DNA is called "junk DNA" by molecular biologists because they cannot see a use for it. Or is there another reason?

Once uncoded DNA is labeled as "junk DNA," the indoctrinationkeyword "junk" is triggered. The result is a pre-programmed assumption that whatever the "junk" may be, it is unpleasant and offers nothing of value, so move along; nothing to see here. And above all else, do not see the special children.

Like the ancient Redwoods, we too have our own deep genetic well, and evolution is already drawing out that which we will need to endure. Those who share our own genetic history now walk among us. They are known as Indigo Children, Crystal Children, and other names. Relatively new phenomena, they began to appear in increasingly greater numbers during the last half century.

This begs the question: What is the triggering mechanism that sets these phenomena in motion? That mechanism is our sun. The Earth changes now underway are principally driven by the sun, and it is changing us as well. One contributing factor is that there has been a detectable increase in overall solar radiation since the early 20th century. That is one contributing factor.

Another is Earth's weakening magnetosphere. The magnetosphere is generated by the iron core of our planet and is about the size of the moon. As the Earth spins, it generates an invisible shield, protecting us from harmful rays. Now, scientists are reporting that it is weakening at a prodigious rate the harbinger of an impending polar reversal where magnetic North flips and becomes magnetic South.

That the weakening is happening is not at issue. It is, and at a concerning rate. However, the question of when this magnetic reversal will actually happen is still up for debate. Yet, we already have these subtle influences at play, a measurable increase in solar radiation compounded by a weakening of Earth's magnetosphere.

Not enough to end all life as we know it, but on the other hand, how much is really needed to activate a genetic trigger? One that initiates a transformative process? This possibility is not science fiction because geneticists have reported that uncoded "junk DNA" can and does change coded DNA. This is what is happening now with these special children, the Indigos.

Named for their indigo-colored auras, these children are awakening in growing numbers. Yet, they are strangers in this strange land of materialism. Brilliant and determined, their values can be very different from their parents, for they will not gravitate toward the rewards of the material path.

Rather, they'll prefer paths of exploration and discovery. For them, the most perfect world is not one of acquisition, but one of harmony. This is why they are drugged, mocked, and maltreated all too often.

Those who succumb endure emotionally painful lives and some are taken early with prescription medications. Others, by chance, are able to develop with the support of their families and they learn to become chameleons.

If you are expecting to see a mindless teenage girl fascinated in boys and fashion magazines, that is what you will see and the facade will be so complete that you'll never suspect otherwise.

However, when these children come to know you and sense your goodness inside, they will reveal themselves to you. When they do, it will be like meeting someone entirely new for the first time. Then you will learn that they are not alone.

Nature has provided them with a small community of mentors, guardians, and elders to shield these precious innocents from harm as they follow their instincts. However, their purpose is not to lead. It is not their instinct to do so, unless by coincidence. Their instinct is to explore the world about them and to join with it in a fulfilling sense of oneness.

Indigos are natural pathfinders and their empathic powers are considerable. When you come across someone new and an Indigo shakes your hand, you've found a new friend or good neighbor. Likewise, if an Indigo avoids someone, so should you.

As an Earth Pilgrim, you'll survive on the go. There will be days when you come to a fork in the road and have no idea of which way to turn. Ask the child. If it takes a day for the child to answer, then rest there for the day and wait. Like a person blind from birth, the Indigo child possesses supernormal powers of perception.

At this point, you may be wondering just what kind of army could be composed of what is essentially a self-appointed Moses, a blind person, a gifted child with an Indigo-colored aura, and a scattergun-toting grandma with a serious game face?

It will be an army of light and you'll enjoy the company.

Where Do You Go From Here?

Making the commitment to be in it for the species is difficult. To be realistic, the sacrifices are immediate and the gains may only be measured long after your years on Earth.

For this reason, you must go slowly in deliberation so that you are fully resolved. Think of these words as the destination for the first of your many journeys as an Earth Pilgrim:

```
"Today, I take personal responsibility for the
survival and freedom of my species and I will do
it alone if need be, and without any expectation
of reward or recognition. This is my covenant with
the Creator to which I bind my own life."
```

If this is not your commitment today, a good first step in this direction is something we can all do as a species to mitigate the severity of what is about to happen.

In a manner of speaking, we must literally reprogram the future by actualizing a power within us all, via-a-vis a global human resonance event.

Yes, we can do this!

12

We Can Reprogram the Future

At first, some will say, "Reprogram the future, sure wish we could!" Others will indignantly scoff at the idea with crass pronouncements such as "What utter nonsense!" In a not too distant tomorrow, however, when everyone is standing in their front yards, pointing up at the sky, and crapping their collective pants, new avenues of thought might become possible, if not desirable.

In other words, do we only look at what is possible when we unexpectedly find ourselves in the midst of a crisis? An example of one such crisis is December 7, 1941, the day Japan struck Pearl Harbor with a devastating sneak attack that plunged America into WWII. But was it really a surprise? Consider this:

In 1924, US Army Gen. Billy Mitchell toured the Orient and upon his return wrote a 323-page report in which he predicted that Japan would attack Pearl Harbor with carrier-based aircraft on December 7, 1941, at exactly 7:30 a.m.

How was Gen. Mitchell's report received? It provoked an army of indignant and outraged naysayers, debunkers, and cynics. Their ire was so great that Mitchell found himself before a court martial and was forced to resign his commission in 1926. Yet, his 1924 prediction was off by just a matter of minutes.

After a span of some 17 years, how many watches could keep time as well? Then, or today for that matter? Yet, for the soldiers and sailors who died on December 7, 1941, the hubris and ego of these self-appointed cynics and debunkers came with a terrible price. We can only imagine the anguish that Mitchell, who passed away in February 1936, would have felt had he lived to see that day.

Ask yourself: If history is to repeat itself, then is this the kind of history we want? Again?

Or, do we listen to our inner voices, no matter how great the derision may become? This is a matter of personal choice and, during the early spring of 2000, I had an experience that was as profound for me as Gen. Mitchell's 1924 tour of the Orient was for him.

For me, it happened during a visit to Israel with my wife, in a different corner of the world far from my California home. During this trip, I experienced an epiphany while standing atop Masada, an ancient desert fortress overlooking the Dead Sea. It was the defining moment that would launch me on my present course.

What Is the Lesson Here?

What drew me to Masada was partly curiosity like any other tourist, but also a deep sense that perhaps something was there that I needed to learn. Thankfully we arrived early as I spent the whole day atop the fortress, eagerly exploring any place one could walk to, crawl into, or stand in. If it was accessible, I was there, yet nothing led me to my goal.

Illustration 63: Marshall Masters (2000)

The knowledge goal called to me, but I could not find it and this was on my mind as I walked along the western edge of the fortress. There, a small group of tourists was listening to a lecture given by a tour guide, and she was captivating—so much so that I snuck up to the back of the group to listen in.

She was a bright young archeology graduate student earning a few extra shekels on the side by guiding tour groups around the fortress, and behind her was the Roman ramp used to breach the fortress in 74 CE, by Roman governor Flavius Silva, who had laid siege to the thousand Jewish zealots living atop the mountaintop fortress the year before. Silva's massive rampart was raised against the western approaches of the fortress with Jewish slaves.

To slow or halt the construction of the ramp, the zealots atop Masada could have easily rained rocks down upon their enslaved brethren. Instead, they chose to save their lives by forfeiting their own.

In the spring of 74 CE, the fortress was finally breached, but for the Romans, it turned into a quintessential Pyrrhic victory. After breaching the defensive wall with their battering ram, the legionnaires stormed the fortress only to find the dead bodies of the defenders. Save for two women, the zealots had burned the fortress and then taken their own lives.

When the lecture was over and after a few announcements the group was freed to pursue independent interests. The guide then turned around and idly strolled to a spot along the western wall, with a spectacular view of the Roman ramp below.

Then my elusive quest piqued my curiosity just one last time and, seeing an opportunity, I walked over to where she was standing and stood quietly beside her.

We'd both been gazing idly at the Roman ramp for about a minute when I commented, "I'm not in your group and I hope you did not mind me listing in, but your lecture was fascinating. Yet, there is one question you did not answer and it should have been asked."

That got a raised eyebrow and a curious, "And what would that be?"

"What's the lesson of Masada?" I asked, "That it is good to die for your beliefs?"

The question made her chuckle and so she turned to me and plainly said, "No. Dying for your beliefs is not what we learned here. What Masada taught us is that you always need to have a Plan B."

Wow! In that instant, I received an epiphany that has been at the core of everything I've sought to do since then. To find a Plan B for humanity. One that could bring us all together in harmonious resonance.

Fast forward a decade to present. I'm writing this book, and once again, like that day on Masada, something needed to be found. But what? As before, I had explored every place one could walk to, crawl into, or stand in, and yet the single point of truth, or SPOT, I needed to connect all the dots eluded me.

Tired and frustrated, my wife suggested a family outing to an exhibit of post-expressionist painters at the de Young Museum in San Francisco. Perhaps a bit of

an explorer geek (which is how I view myself), I was not about to peruse the galleries without an audio tour player strung about my neck. Zigzagging from one painting of interest to the next, I punched numbers into the player and gawked like a good tourist. It was a very pleasant day of idle distraction and then something incredible and unexpected happened.

I was standing in front of a painting by Vincent van Gogh, titled *Starry Night over the Rhone* (September 1888), admiring his use of color to paint a vivid night sky and I noticed an audio tour number and punched it into my player.

The narrator was a woman with a very soft and yet sparkly voice, subtly enhanced with a well-mixed soundtrack. All in all, a well-produced sharing experience where the program showed me the art through van Gogh's eyes, as opposed to dryly telling me about it.

The soft voice explained how this painting is admired for two reasons: van Gogh's use of color and the manner in which he applied the paint to the canvas. The brushstrokes were something I hadn't paid close attention to before, but now, with the guide calling my attention to them, I could see them.

With this new awareness, the whole painting took on a new and powerful meaning for me. As I studied the brushstrokes, I felt the passion that had driven van Gogh to paint this picture. As the audio narration faded away, I stood there taking it all in for a few minutes and turned to see that several other men and women had crowded around behind me to view the painting as well.

Most had audio players like I did; several people were also having the same first-time experience of seeing the brushstrokes. The sparkle in their eyes was testament to that and they too now studied the painting with a shared sense of wonder. Yet, they were all of different ages and walks of life.

In that moment an unexpected epiphany came upon me. The de Young museum was providing a secular, artistic experience for folks like me and those standing about me that connected us to the painting in a rather magical way.

While the genius of van Gough was the opportunity for my epiphany, it sprang forth from the manner in which the museum curators had guided my experience, replete with the right atmosphere, staging, and presentation.

With that epiphany, I had my SPOT. We each have a natural ability to fully appreciate and to be inspired by genius, but only when we understand it. Ergo, you can never be a true genius until people can appreciate your gift.

What the museum had done was help me use my own natural gift from the Creator to see this painting in a very new and empowering way. In other words, an artful presentation of genius is genius of a kind as well. It's all in the presentation, because from there we can each fly on our own.

Do you believe that you too can fly on your own? Or do you view the world at large through blinders, glued to your brain, ostensibly for your own good? That question reminds of something I learned decades ago during a flying lesson.

Night Flying

I had just soloed and was admittedly getting a little cocky. So one day while returning from a local training flight my instructor casually asked, "Do you know what the nighttime emergency procedure is for a single-engine aircraft?" He made a point of looking serious.

Like a dozing student caught flat-footed, I answered, "Huh?"

He cleared his throat and said, "When flying a single-engine airplane at night and your engine dies you first attempt to restart it. If that is not possible, then descend to an altitude of 100 feet above ground level and turn on your landing lights."

Now he had me. "That's it? You just fly down to a 100 feet and turn on the landing lights?"

"Well," he wryly answered, "if you don't like what you see, turn the landing lights off." Oh nuts, I'd been had.

He later explained the point of this humorous anecdote. Unless you absolutely must, flying at night is something you leave to more experienced pilots flying mulch-engine aircraft.

In terms of reprogramming the future, the point of this anecdote is that we're not single-engine pudnockers (small, slow private airplanes). We do not need to fear the night because someone else has something we're led to believe we do not possess ourselves.

When it comes to the possibility that we can reprogram the future, do not let anyone talk you into thinking and behaving like a lowly, single-engine pudnocker. This is because when it comes to making choices about the future, each of us is a hot-shot pilot with a mulch-engine aircraft.

Therefore, to reawaken our inborn talents, we just need to rack up more flight time. In a manner of speaking, we punch the right numbers into our audio tour players of consciousness. And then we fly—together.

In this case, we use a global resonance event to connect the greatest number of people from around the world to a powerful purpose: Reprogram the future through intention. So, is this Pollyanaish New Age thinking, a noble question with form yet lacking in substance?

To answer that question, we will look at something that already benefits us each time we fly in a modern passenger jet. All new designs are first flown in simulators long before the first rivet is fastened. With this in mind, let's imagine that we're running a simulation of a global resonance event called Project EarthSave48. It will be an event in which we use the technologies, skills, and natural abilities we already possess to reprogram the future.

The Project EarthSave48 Simulator

The critical technology for this simulation is well proven by the amazing outgrowth of online intention groups worldwide. They are using the World Wide Web to cross all boundaries and seas with globally coordinated prayer and meditation events—each with its own set of goals, methods, and intentions.

Building on this, Project EarthSave48 uses this familiar approach, but with a significantly enhanced presentation experience—one that is possible, but not yet affordable for present-day, Internet-organized prayer and meditation events.

However, this simulator will be infused with resources and talents stemming from the global realization of a clear and present danger to all life on the planet. With survival as a compelling new imperative, money becomes no object.

Even if the remotest chance exists that the future can be reprogrammed in a way that mitigates the disastrous consequences of a Planet X flyby, it will be a Plan B that people can accept.

Plan B will not conflict with their current Plan A beliefs, which range from divine intercession to science-saves-all and everything in between, all of which are perpetually irreconcilable. The best that can be said of the world's Plan As is that there are those who learn to live with a bad marriage and those who massacre half the family.

Because Project EarthSave48 is a secular, Internet-based Plan B, it avoids the quarrelsome and violent quagmire of religion altogether. It is about individual choices. Those who wish to remain firmly planted in their own Plan A can organize their own global event using the concepts developed in this effort.

For those wishing to participate, they will be joining in a global effort in an Internet project with four key features:

- It is a non-denominational and secular global effort for the widest possible inclusion.
- The intention event creates a cohesive and transformative experience for all participants.
- It uses measurable scientific goals with a heuristic, trial-by-error process.
- It is the first ever use of Internet intention vortices to reprogram the future.

This will be a bold and ambitious project, on a scale never before seen in the annals of human history and, like a new passenger jet design, we can run simulations to see how something like this could be organized, staged, and performed to change the future.

Therefore, in this final chapter, you will witness a simulation of the very first Project EasthSave48 intention event to spare our world from as much harm as we possibly can by using the power of human intention to blunt the effects of a Planet X flyby.

In examining the feasibility of the Project EarthSave48 concept, this chapter will first present the theoretical foundation for this first-of-a-kind global effort. This will provide the necessary context for a simulation of the first of these global resonance events in the second part.

So then, where do we start?

There is an old business adage that best defines the necessary starting point here: Nothing happens until somebody makes a sale. In other words, the sale is the trigger event from which all other efforts are organized. The same applies here. What could these trigger events be?

The Trigger Event

A project with a massive global scope such as EarthSave48 will need a significant trigger event to overcome the pronouncements of the can'tologists (i.e., those who say you can't do this, you can't do that, and you can't do the other thing until I say you can).

What could a trigger event look like? Something that leaves us feeling helpless and in the crosshairs of worse to come? Perhaps, a stunning global catastrophe, larger than Sumatra (2004), Katrina (2005), and Haiti (2010) put together, such as a major asteroid impact event?

Whatever the trigger event turns out to be, one outcome is certain. The all-knowing pronouncements of the self-inflated can'tologists will begin to fall on deaf ears. People will be ready to take chances and to invest heavily in them.

For this reason, a project like EarthSave48 will be an attractive gamble to a large minority of the global population because of the five following organizational goals:

1. A nondenominational, secular approach.
2. Immediate utilization of current human and technical resources.
3. Tested technologies with a preference for open-source solutions.
4. Lowest possible participant learning curve for participants.
5. Participatory internationalization for all participant languages.

The first four are concepts common to all business and organizational planners. However, language support through a participatory internationalization effort is starkly different to how most computer software and Internet companies approach their internationalization requirements. Therefore, it requires close attention.

Intuitive to All Languages

Today, the language of software is English, but the documentation for that software must serve the needs of users in multiple languages. Consequently, most firms today have their documentation written in their native language and then translated by third-party contractors at a per-word rate. They can also run what are known as "machine translations," where translation software does the bulk of the translation followed by light proofreading by junior editors.

The result of these lengthy, low-bid, two-stage approaches is translated documentation that is grammatically precise, but all too often obscure and confusing. This is because the context that a native-language author can provide is deemed to be too costly.

However, for Project EarthSave48, internationalization will be based on an English-language core. However, instead of the after-the-fact cheap translations used today, these translations will be done in real time with participating authors. This way, vital information can be passed on to all volunteers and participants in a quick and efficient manner that produces results that are consistent and comfortably intuitive for all participant languages.

Work performed at the core of the internationalization documentation effort will be in English, with all other languages represented in a collective author pool. To make this effective, each unique language author or volunteer must be fully fluent in English as a second language. This way, they can participate in the analysis process by contributing their thoughts and ideas.

Once a final version is ready, all represented language variants can quickly adjust their own work-in-process to reflect the English core effort. Given the scope of intention for this project, this internationalization approach is of paramount importance, especially given the galactic scope of the EarthSave48 intention goal.

Galactic Intention

The supernormal proposition that defines the scope of this project is as follows: Intention is a conscious thought that sets all things in motion and in a manner organized by the universe itself—something best described, in this case, as galactic intention.

No doubt can'tologists will tell us we don't grasp all of what is required by such an ambitious proposal; however, if the truth be known, it is they who lack the grasp. This is because it is no more necessary for us to understand the mechanics of how the universe organizes our intentions than it is for us to understand the inner workings of the automatic transmissions in our cars.

Even the Creator agrees; simple works best. All we need to know is how to shift our intention into "D" for drive and let things work out for the best. And here is the good news. Thanks to the World Wide Web, humankind has a monster truck-sized "D" for drive and it works for the whole globe.

How do we know? Simple-works-best solutions are growing in number, scope, and scale and never before have we seen so many organized prayer and meditation events. Furthermore, the trends are ticking upward like rockets.

Plus, the intention of these global prayer and meditation events is incredibly diverse; they can include such things as world peace, relief from a natural disaster such as a drought, healing for the ill, and expressions of love and healing for Mother Earth herself, just to name just a few.

While these various intentions and methods serve noble, yet divergent goals, they all have one thing in common. They are all Earth-centric. In other words, they are only directed to effect a change in someone or something on this planet. None of these events has ever been organized to alter the orbital behaviors of a near-Earth object, which is the very goal of Project EarthSave48.

This is not to diminish the results of present Earth-centric intention events, but to recognize their generalized approach. Earth-centric events rely on a shotgun approach, as marketing gurus describe it. This is an approach that is much like throwing spaghetti on the wall to see what sticks. Inevitably something will, but what exactly? This is why measuring the success of these intention events can be difficult, at best, for uninvolved outside observers.

What outside observers need in this case and what is absolutely vital to the prospect of a successful effort is something marketing gurus call a "rifled" approach—one that is precise and can yield clearly measurable results against a near-Earth object in space by means of a galactic intention.

Measurable Scientific Goals

The need for measurable scientific goals is why Project EarthSave48 offers a first-ever hybrid, a seamless fusion of scientific goals with terrestrial and galactic intentions, to produce clearly measurable results on a galactic scale against a near-Earth object in space.

This is not to say that scientists get a free ride when it comes to sharing the risk of failure. They have to create the aiming solutions for these events as well as a way to measure outcomes. These scientifically defined event goals must aim the rifle of a global effort which brings our fullest supernormal abilities to bear on a specific leverage point in space.

For example, when a space threat possibility looms, the instinctive assumption is, "Let's nuke it." The assumption is based on the known destructive power of nuclear attacks on cities and their large populations. However, in space, this Earth-centric assumption fails to factor in the behavior of large, fast-moving objects in the vacuum of space.

Consequently, nuking a large asteroid is likely to turn one big ugly rock into a hellacious storm of smaller ugly rocks. Scale this reality up to something like a planet-sized object or bigger and shooting it with nukes would be like hunting a hippopotamus with a BB gun. You're just not going to stop the beast from doing what it intends to do.

Therefore, can we use a galactic intention to destroy Planet X? Not likely at this stage in our evolution. However, if we aim what we do have to create a little bit of leverage in just the right way and at just the right place, we'll be within the realm of something we can do today. The keyword here is leverage.

Leveraging Galactic Intention

When young boys and girls are introduced to the martial arts, such as judo, they learn early on about leverage. Through careful practice, in time they gain the confidence and skill to throw someone twice their size, by leveraging their opponent's mass. The same strategy applies here.

During the 18-month channeling study, we were often told the following about Planet X:

- The description often given by the guides (entities) is of a mini-constellation with several satellites and there is nothing we can do to prevent a flyby through the core of our system.

- Planet X, like comets, is a fast mover and its comet-like orbit is steeply inclined to the ecliptic and therefore unpredictable. Furthermore, the guides tell that it is unstable in both its vertical and horizontal axes. This presents us with a unique advantage as well as a terrible risk.

- If the orbit of one of Planet X's satellites were to change dramatically, say through a collision, this could significantly change the effects of a flyby for the better or worse.

- We as a species possess the ability to form a global coherent bond of spiritual power to focus a powerful intention against one of Planet X's satellites. By moving that satellite's orbit, we can then effect small orbital changes in the other bodies within that mini-constellation, pushing it into a more benign orbit relative to Earth.

Simply stated, the guides have repeatedly told us that we presently possess the natural ability to mitigate the effects of this flyby, provided we do it on a sufficient scale. If this sounds dubious, consider the dwarf planet Pluto and the biggest of its three moons, Charon. Closest to Pluto, Charon is the largest moon in the solar system, proportionally speaking. This is why it does not orbit Pluto in the manner that our own moon orbits Earth.

Unlike Earth and its moon, Pluto and Charon orbit a point in space between them, called a center of gravity. This is because Charon is proportionally large relative to Pluto. Consequently, the two orbit each other in a way that mimics two chrome beaters whirring about each other in the mixing bowl of a kitchen stand mixer.

The point here is that we can mitigate the effects of a Planet X flyby by altering the orbit of one of its satellites, which in turn will exercise a bit of leverage to cause this highly unstable planet to deviate from its present orbit into one that more favors Earth. In other words, we help save our planet with leverage in a cosmic game of pool. Something that all whitebelts in judo learn the first day they throw someone twice their size. It's the leverage!

Assuming scientists and astronomers can come through with a good way to aim our cue ball, what do we use for a cue stick in this cosmic game of galactic eight-ball? It is something I call an intention vortex. Like a cue stick, it will be the force amplifier that sets things in motion.

Intention Vortices

When you see water spinning around in a bathtub drain, you're looking at a small vortex. Likewise, when you see a massive hurricane like Katrina (2005) headed for land, you're also seeing vortex, but on a terrifying scale. Either way, something is going down the drain.

To create an intention vortex, the first thing that needs to happen is the scientific formulation of an intention event goal. Once the goal is formulated, responsibility for the actual event passes out of the hands of the scientists and to the spiritual volunteers and participants. The participants will play two interchangeable roles: aimers and energizers.

Organizing the intention vortex begins with a city, or passenger ship at sea, selected by the science advisers as a favorable intention vortex site. The inhabitants or passengers at these sites will then serve as the aimers within the eye of the event itself, as the center of the vortex.

From within the eye of the event, the intention vortex is aimed at a target in space and, with the help of animation graphics, all participants will eventually look skyward through the eye of the intention vortex, much like the eye of a hurricane, out into space.

A deep understanding the mechanics of the event is not necessary and would probably be counterproductive. This is why the participants need to be sensitive to the spiritual energy flowing from them, to them, and around them from all the other participants in the world, the energizers. Energizers will do this by directing their collective intention into the base of the vortex.

To help understand how it is possible to direct all global psychic power to a single point and then to redirect it out into space, let's revisit our example of a game of eight-ball to compare complexity vs. resonance.

A Game of Cosmic Eight-Ball

Assume you're out for the night with three friends for a friendly game of cosmic eight-ball at the local pool hall. You're well into the game and now you need to put the five-ball in a corner pocket. However you do not have a straight line of sight from the cue ball, through the five-ball, and on to the corner pocket. Consequently, you know that you need to angle your shot so that the cue ball glances off the five-ball in a way that achieves the desired result.

To make the game interesting, you all decide to conduct two experiments that evening. In the first experiment you want to test the classic shotgun approach of the type used by present-day Earth-centric intention events.

The goal here is to see if many such smaller events conducted simultaneously across the world could be as effective as a single, coordinated intention vortex event. Therefore, in setting up this first experiment, the four of you decide to all aim each your sticks at the cue ball. Then, everyone does their best to simultaneously hit the cue ball. So, will complexity be a limiting factor here?

If we add the corner pocket, five-ball, cue ball, and stick together for a baseline complexity factor of 4, by what measure will that factor increase as each player lines up for the shot.

In this case, the increase in complexity is not expressed with simple addition (i.e., 4, 8, 12, and 16, respectively). Rather, it is squared with each additional player (i.e., 4, 16, 256, and 65,536, respectively.) As the differences here are rather dramatic, we need to use a galactic intention scale, as opposed to a more confined Earth-centric scale.

In a manner of speaking, Planet X is the biggest five-ball we've ever seen and its satellites will be a swarm of smaller cue balls orbiting about it. Consequently, this gives us a good clue as to how difficult it will be for scientists and astronomers to compute a targeting solution for the intention event. It will require an effort that will use computer resources millions of times more powerful than Deep Blue, the chess-playing computer developed by IBM to defeat World Chess Champion Garry Kasparov in 1997.

Therefore, by using a decentralized shotgun approach where every city or group does it own Earth-centric thing against the same large object in space, complexities from both sides of the effort will be a severely limiting factor.

Conversely, a rifled approach used with intention vortices will not only slash the level of complexity, it will also offer one significant benefit—coherent resonance that is focused, uniform, and therefore measurable. But can an intention vortex deliver the desired results?

To test this assumption, you and the other three players decide to conduct the second experiment.

As before, you set up the same five-ball in the corner shot, except this time only one pool stick is used. Yours. Once you've aimed your shot, the other three players line up behind you and gently hold on to your stick with one hand.

When it comes time to take the shot, you call the cadence so that you all move with a single choreographed motion as you collectively hit the cue ball. You know the more coordinated your efforts are, the greater the result will be. Therefore, you practice the stroke with your friends until it becomes so well synchronized that it resonates.

What this means is that each player no longer needs to be told what to do. Each one has trained himself to sense what the others are doing, and all that is needed is a simple cue to set everyone in motion. This being achieved, the co-originated intention of striking the cue ball is dramatically amplified.

So amplified that not only do you sink the five-ball in the corner pocket, but your collective effort has also sent the cue ball flying out into space. Moments later NASA ground controllers receive a mysterious message from the International Space Station "Houston, we've just had an encounter of the third kind with a high-velocity cue ball." OK, granted, that part goes to the extreme.

Nonetheless, we can now see how the resonance possible with a coordinated global projection against a single aiming point serves two useful goals. It lowers complexity in the overall effort to help scientists and astronomers plot a targeting solution, while also magnifying the force of the collective effort through synchronized resonance.

For this reason, placing the eye of the intention vortex, the aiming point if you will, is where the first player sets up the shot. Then all other participants energize the intention vortex for highly amplified results.

All participants must therefore understand and agree that by "doing their own thing" they will hurt the overall effort. This is because consistency is a prerequisite for the reliable measuring of results obtained through a heuristic, learn-as-you-go effort.

The first step in this direction is orientation. Participants must be able to easily visualize in their minds the spot wherever they happen to be standing relative to the aiming point of intention vortex.

This is done with a simple Project EarthSave48 participation orientation system comprising 48 global event zones.

EarthSave48 Global Event Zones

The scope of effort for Project EarthSave48 is to project a focused global intention at a closely defined area in space. On the scientific side of the effort, the program relies on precise measuring systems to create an aiming solution for the global effort.

On the other hand, the orientation needs of participants and volunteers do not require the complexity of these measuring systems. For energizers worldwide and outside the projection zone, all that is needed is an intuitive way for each to fix their own global position relative to wherever the intention vortex is to be created. To do this, we begin with concepts that are most commonly known and understood from all around the globe.

Our planet is divided into two hemispheres (Northern and Southern) and 24 time zones that give us a standardized way to tell the time in any part of the world. By combining these popular concepts, we have a simple global orientation system for Project EarthSave48 participants and volunteers.

To do this, we multiply 24 time zones by two hemispheres, yielding 48 discrete event zones numbered 00N through 23N (in the Northern Hemisphere) and 00S through 23S (in the Southern Hemisphere).

How does this system compare with more precise global position concepts? To answer that question in terms of the layperson, let's use an example where Tokyo, Japan, is the projection zone.

System	Tokyo, Japan	Description
Global Map Coordinates	35°42'2" N 139°42'54" E	Difficult for the layperson, it is the most precise way for cartographers, navigators, and pilots to pinpoint Tokyo, Japan, on the globe.
Time Zone Offset	UTC +9 Japan Standard Time (JST)	Midnight in Greenwich, England, is 9 a.m. in Tokyo, Japan, and everywhere else in both hemispheres for JST. Actual times vary with country daylight savings time preferences.
Project EarthSave48 Event Zone	09N	Tokyo, Japan, is in the ninth time zone to the east of Greenwich, England, and it is in the Northern Hemisphere (N). Ergo, 09N.

Granted, these 48 event zones lack the specificity of more precise instruments, but they give all participants an easy and uniform way to visualize where they are presently standing upon the Earth relative to the projection zone. To do, all that is needed is a time piece with a 24-hour face, or a simple hand-drawn diagram for that matter.

In a functional sense, one could compare the purpose of these event zones to a common observatory dome. When we watch television programs related to astronomy, we're invariably shown massive doors opened on the dome to reveal the telescope. As the dome is rotated to the desired view of the sky, the telescope is pointed through the narrow wedge created by the open doors.

To visualize this, imagine that you are standing in the field where the Avebury 2008 formation (00N) first appeared and the projection zone for the intention vortex is Tokyo, Japan (09N). Like the massive observatory dome, you rotate to face the wedge between the retracted doors. Then like the astronomers operating the telescope inside the dome, you can fine-tune your intention in resonance with others, much the same way astronomers do when focusing on a specific point in the night sky.

However, there is one critical distinction here between a spiritual, global resonance event and the manner in which astronomers study the night sky. It is called line of sight.

For astronomers, the line-of-sight limitation is absolute. They must have a clear view of whatever they're looking at through their telescopes. This is why large observatories are typically found on remote mountaintops, far from the light pollution of large cities and elevated above most natural distortions as well as those caused by manmade pollution.

However, with a spiritual event such as an intention vortex, there is no such physical line-of-sight limitation. This is because the event is not defined by what is seen through a telescope, but what is seen through the unrestrained power of human imagination.

Therefore, in the spiritual sense, our intentions are not projected in a line of sight, but rather in a line of visualization. Nonetheless, the science teams who prepare aiming solutions for an intention event must provide participants with a simple way to visualize the elevation of their focus.

As with the 24-hour clock face used to help volunteers and participants orient themselves toward the projection zone, we can likewise use a 12-hour clock face to fine-tune that view. We do this by adding a simple reference system similar to those used by aviators with a 12-hour clock face view, where 12 o'clock high is directly above our heads and 6 o'clock low is directly beneath us.

For example, let's assume you are standing in the Avebury 2008 formation during a visualization; you're in zone 00N at latitude 51° whereas Tokyo is in 09N at latitude 35°. Obviously, the viewing elevation will be different depending on the latitude.

For example, energizer participants in one city will need to look up with a 10 o'clock high elevation, whereas energizer participants in other cities at other latitudes will have to fine-tune the elevation of their view so that everyone is focusing on the same projection zone. The result is a very focused and highly amplified invention vortex.

For participants the Project EarthSave48 orientation system will be an easy skill to quickly master. However, the degree to which they actually contribute to the overall effort will be determined by their own ability to project their intentions within the framework of a coherent resonance effort.

This begs a natural question. If personal ability is the overarching measure of global potential, who comprises the most optimal pool of participants?

The Optimal Participants

The optimal participant pool for the Project EarthSave48 consists of young children in the company of their parents. Simply put, a young child is best suited to projecting the type of intention required for this project, because children are not burdened by years layered with disappointments.

Conversely, adults who choose the path of the light worker must devote a great deal of time and effort to stripping away these very same layers as a necessary first step in the pursuit of wisdom. This will help them develop more advanced techniques to amplify the natural abilities expressed by young children. Hence, the popular term "seeing the world through the eyes of a child."

Although light workers represent a valuable pool of participants for a global resonance event such as an intention vortex, they lack the numbers needed to be deemed optimal. However, as volunteers in a role of leading intention visualizations, our available pool of light workers is both optimal and necessary. In this case, they will focus their principal efforts on working with young, participating families.

Young families are truly optimal in two ways: support and interaction. Support for many young families will come through their networks of families and friends. In terms of interaction, parents of young families routinely see the world through the eyes of their children with family films and teaching programs. For example, when parents sing along with their children in response to cheerful antics of Count von Count on "Sesame Street" and other such television characters, they are focusing with the child on a specific intention—that being to understand what Count von Count is counting.

In such a simple and life-affirming thing as this, we find the true galactic magic of what can actually be achieved during an intention vortex. Herein we find the necessary manner in which these young families are prepared as participants

Preparing the Families

We use what already works—family venues like Disneyland, "Sesame Street," "The Muppets," and so forth.

Families are taught in a Sesame Street-like style, through the eyes of their own children. Through this shared learning experience and with proficiency developed over time, these families become true intention powerhouses, especially when they combine their efforts with other families sharing similar interests.

However, a prerequisite for all participants is that they must agree to follow the program faithfully and in detail. This effort is to create a conscious thought based on science that sets things in motion and in a manner organized by the universe itself.

This is not a one- but a two-stage decision.

The first stage is to agree to follow a secular, non-denominational approach that allows for a measurable, heuristic process—one that is subject to purely human, trial and error with a conservative, do-no-harm approach to the possibility of failure; and for the following three reasons:

> 1. If this project is viewed as a global prayer for intercession by a divine being, the pivotal role of science in the success of the effort will be stained by a world view that is generally untenable for many scientists.

> 2. Should the project be co-opted into a global prayer for intercession, it will fail for the same reasons that keep the world's major religions perpetually in conflict with each other.

> 3. A global prayer for intercession negates a principal benefit of this project —that we accept the responsibility to become our own messiahs.

Those who cannot participate on these terms are encouraged to work independently with like-minded others to organize their own intention events and in a manner purely of their own choosing. Given that current Earth-centric mass prayer and meditation events employ a well-understood "shotgun" approach, they'll have a broad range of proven event templates to choose from.

This brings us now to those who are willing to participate on the specified terms. The second stage of conceptual commitment is as difficult for some as the first. It involves a commitment to self-perception as a participant, as defined by:

- Intention originates in the mind, which exists outside the physical brain.
- Our physical brains are receivers of consciousness, much like the radios in our cars.
- Our mind-brain connection is throttled by self-governing limitations. When these artificial limitations are removed, we can fully tap into the mind-over-matter potential of the human spirit.

This second commitment is vital and here is where the secular requirements may be deemed offensive to the fundamentalist sensitivities of various religious faiths. Likewise, acceptance of ourselves as cosmic beings will no doubt offend fundamentalist thinkers in the scientific community as well. Ergo, it is a given that Project EarthSave48 is not going to be everyone's cup of tea.

In a manner of speaking, what this all boils down to is illustrated by The Story of Goldilocks and the Three Bears. Some will find the porridge too hot and some too cold. However, for those who say, "Ahhh, this porridge is just right," the benefits of this project are substantial.

Those who can make both commitments will embark upon the effort with a realistic view. They'll know that whether the effort has a successful material outcome or not, the event itself will be transformative on a personal level. All participants will experience a substantial increase in their state of awareness as well as in their own powers of observation and contemplation. All these are vital survival skills.

Ergo, there are no downsides to this project for participants who accept the project requirements and guidelines, and the more prepared the participant is, the greater the benefits become.

For humanity as a whole, material success is the aim but if the only result is that young families are better equipped to deal with the hardships of the tribulation, that's good too. They are, and always will be, our best first hope as a species.

Yet there will be the self-interested taunts of cynics and can'tologists, such as "After investing a lot of valuable time, money, and effort into helping young families play Count von Count with Planet X, all we'll get for it is a zero sum result. So why embark upon this fairy tale in the first place?"

What do these cynics and can'tologists have to offer? It is best summed up by the final stanza of "The Hollow Men" (1925), the most quoted of all the poems by T. S. Eliot:

```
    This is the way the world ends
    This is the way the world ends
    This is the way the world ends
    Not with a bang but a whimper.
```

"The Hollow Men" was an expression of Eliot's concerns over post-War Europe under the Treaty of Versailles that ended WWI. It was a misguided treaty that he and like-minded others knew would only serve to fuel the next world war, which is exactly what it did.

This brings us back to that cynical and can'tologist taunt, "After investing a lot of valuable time, money, and effort into helping young families play Count von Count with Planet X, all we'll get for it is a zero sum result. So why embark upon this fairy tale in the first place?"

Here is an answer: We choose to go out, not with a whimper but with a bang. It is our message as a species to you, our most precious survivor pool. May you be all that we ever were, so live for our memory as well as those before us, and know that you are loved.

As the author, please let me be frank with you. If I had any reservations in my mind about this project working, I would not have written the chapter. With every fiber of my being, I know that we can do this, if we smartly plan the world and work the plan. This is where those who choose to volunteer and contribute will be the lubricant that keeps this spiritual engine turning.

Project Volunteers and Contributors

Once a clear and present danger is perceived on a global level, one response would be to immediately capitalize on a global pool of individuals, organizations, and businesses with talents and resources for a global effort of this scale.

Many individuals and leaders will instinctively step forward to make a difference. Their numbers will be small at first, but in time, they'll be joined by people inspired by their personal courage. Furthermore, given the time constraints, the overall management structure of the project must be as flat as possible.

Making the project possible requires an equal infusion of resources and promotion from contributors and volunteers alike. Any attempt to create a vertical, ivory-tower hierarchy will be no more successful than the Soviet command economy of years past. Rather, the project is always localized as much as possible, where volunteers are functionally grouped.

The creation of these groups must be through an organic process where those with something to give can find their own little corner of the project and the camaraderie of like-minded others.

Such groups could address the following roles:

- Project administrators. Program administrators will maintain message board systems or portals for localized volunteer and contributor groups. As the project grows in scale, local contributions and volunteer administrators will be required for each language supported.

- Project ambassadors. Public figures and local leaders will use their celebrity to help create awareness to foster positive awareness for the project and to encourage participation.

- Science and event goals. Support from scientific universities in developing the aiming solutions for each intention event will be necessary. Event goals will be introduced to participants by media-savvy popularizers of science.

- Network support. A global top priority for network bandwidth will be assigned to the project. If Internet service providers are unable to assist directly, large firms can donate the use of their virtual private networks (VPNs). In any case, security will be critical as attacks by violent fundamentalists, malcontents, and cranks must be anticipated.

- Streaming media aggregation. Online firms specializing in online media-streaming technologies will set the Internet stage. They'll host localized media and work with regional network administrators to develop bandwidth load-balancing strategies with an uptime goal of .9999%.

- Content localization and staging. Local museums will serve as the local coordinators for arbitrating and judging content suitability based on local language and customs and control presentation. Locally, they will have final say on staging and presentation.

- Local television production. Local TV stations, independent production companies, and cable casters will be essential in providing localized support for local live production.

- Craft services. Caterers, restaurants, and craft services will be needed to organize food services for on-site volunteers and participants during events. Fuel the people and you fuel the intention.

- Participant guides. Participants are the project's boots on the ground. They will help organize and train groups of event participants and serve as liaisons with other volunteer groups.

- Personal development. Prior to an event, participants will enjoy open access to professionally produced audio meditations, created expressly for the project by volunteer experts in the field of therapeutic meditation and production service. The goal of these programs will be to help participants independently amplify their own natural abilities.

- Technical support. Each language supported will need a localized support group. Ideally, each local support group will serve as a single point of contact for all technical support needs.

- Training and documentation. Each volunteer and participant language will need training and documentation support that is fully intuitive to native speakers.

- Standards group. The overall goal of the project is to rely on open-source and open-standard solutions to create standards. In a similar manner, Project EarthSave48 will need to create its own standards as well to keep everyone singing from the same page, as the old saying goes.

Once local volunteer and contributor roles are defined, those with an interest in becoming involved will naturally organize themselves by interest into small groups. Within these groups, leaders must emerge through consensus. The project will never appoint any leader.

With that thought in mind, we have now completed our examination of the theoretical foundation for this first-of-a-kind global effort. Now, it is time to run a simulation to visualize how an intention event would actually play itself out.

First Intention Vortex Event Simulation

Every organized intention vortex effort begins with volunteer science groups who specify aiming solutions for designated projection zones. Likewise, for this simulation, the first question must be: Who will step up and take technical responsibility for the first event? Further assuming one of the world's space agencies will volunteer to bear this burden, which one of the following is the most likely?

- National Aeronautics and Space Administration (NASA)
- European Space Agency (ESA)
- Russian Federal Space Agency (Roscosmos)
- Japan Aerospace Exploration Agency (JAXA)
- China National Space Administration (CNSA)

In this scenario, I believe the first space agency willing to step forward will be JAXA, and because of a personal observation. It has been my experience over the years that of all the countries in the world with an interest in this topic, Japan is the most open-minded and tolerant. For me, Japan sets the bar.

Unlike Western countries where caustic attacks and disinformation abound, the Japanese are quite remarkable in their view of the topic. In general, they either take an interest in it or politely pass it by. Consequently, Japanese with a genuine interest in the topic are generally better tolerated than their counterparts in the West.

Trusted sources also tell me that the government of Japan is more practical about this topic than Western counterparts and for good reason. Areas such as this are especially vulnerable to the kinds of natural cataclysms that are certain to accompany a Planet X flyby. This is because Japan is perched in the rather unenviable location alongside the Pacific Ring of Fire, which borders this island nation's eastern shore with a nearly continuous line of parallel, sub-sea subduction zones.

With this in mind, the scenario of this simulation will proceed on the assumption that the government of Japan has ordered its JAXA space agency to take a lead role in creating the aiming solution for the world's first intention vortex event. The city chosen for the first projection zone is Tokyo, Japan, in event zone 09N.

JAXA science teams are then given 30 days to formulate their aiming solution, using their own resources plus whatever other government and private-sector assistance is offered.

Meanwhile, local museums in Tokyo and elsewhere in the world immediately shift into full gear to organize local aimer and energizer participant sites.

In support of the preparation effort, broadcast television and cable networks throughout Japan begin to interrupt their schedules to focus on public education as to the basic workings of Project EarthSave48.

Using live feeds from Japan, networks from around the world begin reporting on preparations throughout Japan and elsewhere, as similar participant organization efforts occur worldwide while intention energizer sites spring up like desert flowers after a good rain.

For the purpose of this simulation, our energizer site will be the stern of a large ocean liner, anchored off Kona, the western coast of the big island of Hawaii. Incidentally, the state of Hawaii USA has the second largest population of Japanese Americans in America.

In Tokyo and Hawaii and elsewhere in the world, participation in the event is rapid, enthusiastic, heartfelt, and dramatic. Organizational tasks that would normally require large amounts of time and effort are given top priory as local museums begin enrolling participants in the event.

Grouped according to preference, participants are matched with volunteer participant guides. Their guides then tutor them using event training materials and introduce them to whatever personal development programs are currently available.

Planners from local theater houses, auditoriums, and sports arenas will use staging guidelines developed by local museums to organize their sites. This includes installation of three-screen audience displays and high-fidelity, surround-sound audio systems, plus coordination with craft services, personal conveniences, and so forth.

As technical volunteers go about the business of preparing local venues, local television stations, production houses, and other technical volunteers install and test transmission systems for live feeds (one to downstream media to their audience displays and sound systems and another to upstream a live production feed of their own participants) to various global aggregators.

Anticipation is building and three days before the scheduled date, the JAXA agency and top government officials hold a long-awaited, global press conference. They announce that JAXA, with the help of other nations, has developed an aiming solution with the intention vortex and that the event is to begin at 9 p.m. Tokyo time.

What follows are random samples of sound bites and snippets, taken from the ensuing media blitz:

> "The islands have been abuzz with preparations since the beginning and everyone is anxious to hear... wait... this just in. Tokyo is on for 9 p.m. their time, which will be 2 a.m. for most of us here in Hawaii."

> "I'm live from San Francisco and every ship with a horn is tooting it. Oh my God, I hope you can hear me. The announcement was, was just electric... just knowing and I'm told that cruise ships all over the world are sounding their horns to celebrate the announcement and lavishing their passengers with champagne cocktails."

> "After careful study, the lead JAXA science team decided to reduce the orbit of the second-largest satellite in orbit around Planet X, based on a massive data modeling study using the largest computing clouds ever in the history of the world. The JAXA science team spokesperson also noted that while there were more ambitions solutions with possibly greater yields, this does no harm, and the aiming solution was generally regarded by all involved as the least risky."

> "When asked if they felt the JAXA targeting solution announced yesterday was perhaps a bit weak-kneed, as some have called it, Virgin Galacticastronauts and their CEO enthusiastically applauded JAXA's decision. They also joined in pointing out that this first effort offers a good test bed, with measurable results. Consequently, the lessons learned during first initial effort will enable more ambitious goals in future events."

"This just in. In what is seen as a very risky gamble for the Imperial Family, the Emperor and Empress of Japan just announced that they and the entire Imperial Family have completed their training as event participants and look forward to joining with others in this momentous occasion. The whole nation is abuzz and excited, and this momentous announcement has spawned a tsunami of fresh hope. It has melted away the cynicism of naysayers and energized a new public sense of what is possible."

"In Tokyo, busloads of orphans begin to arrive today at the Imperial Palace, a large park-like area. An initiative of the Empress, this effort is coordinated with museums throughout the country. These children were specially chosen to participate with the Royal Family. Meanwhile, participants throughout the country are making final preparations and cities everywhere are preparing celebrations for afterward, many with fireworks."

On the day of the event, the eyes of the world are on Tokyo as families and friends in every corner of the globe come together in participation sites and elsewhere. As the sun sets in Tokyo, the normal hustle and bustle of everyday life everywhere in the world is mostly quiet as the world draws a collective breath.

Tokyo: The Night of the Event

In venues throughout Tokyo and the world, staging volunteers and audio video technicians double-check their equipment and go over their systems one more time, as craft service volunteers prepare warm and comforting snacks for the participants for before the event and celebratory banquets for afterward.

All throughout the day, a commercial-free continuous marathon of famous singers and performers has filled the airwaves with noble songs and messages of love and hope for each other and our world. In Japan as elsewhere, families, neighbors, and friends gather to experience the event as it is broadcast live on local television channels and via the Internet.

At 7 p.m. Tokyo time, the aimer participants gather in front of their venues and meet their participant guides at pre-designated meeting places. They are escorted to their preassigned seating areas, where they meditate, enjoy snacks, and talk quietly among themselves.

In Kona, Hawaii, on the western side of the big island, all of the large hotels dotting the coast are filled to capacity with celebrants from all over the world. Offshore, a brightly lit flotilla of cruise ships is anchored.

Everywhere in the world, there is a new and growing sense of connection and consciousness between millions of people—that as a species we are becoming one in spirit, for a noble aim. Like the faintest possible whisper, it is there for many and far enough above the threshold to be real.

Grandparents and seniors everywhere remember the last time humanity shared a global experience such as this and recount where they were and what they were doing in 1969, on the day astronauts Neil Armstrong and Buzz Aldrin (Apollo 11) became the first men to set foot on the moon.

They'll also remember Armstrong's first step on the lunar surface and how witnessing that moment changed their world view of what is possible. In the same breath, they'll also share the feeling that like 1969, this could be another "giant leap for mankind"—the harbinger of a sentient species on the verge of a stellar evolution.

Video directors working with the science teams present their current video animations cut for the intention vortex to be sure they correctly present the most current JAXA aiming solution update. The production systems used by the top animation studios in Hollywood to generate animations are designed to incorporate any last-minute change.

This way, minor changes are easily factored in and the animations are quickly rendered for final review and approval by science and intention meditation experts. Once approved, regional data centers all over the world prepare localized animations for automatic push updates to local event venue media centers.

In the venues of Tokyo, participants are settled in and ready for the event to begin. Many will be in rectangular venues which could be staged as shown in the example below.

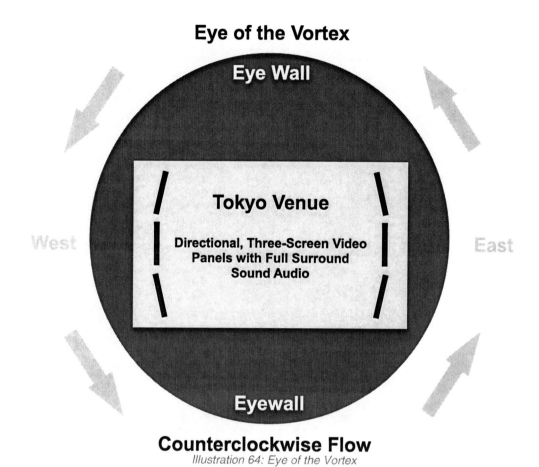

Illustration 64: Eye of the Vortex

At both ends of the room are identical three-panel wall displays. Each panel views outward through the eye wall of the vortex and those standing in the middle of the room can face in either direction for a 180° view.

The support media for the projection zone venues in Tokyo will offer participants two different points of view. The first, in which they look ahead through the eye wall of the vortex, must the same as the eye wall of a hurricane. However, in this case, the overall color will be a golden hue.

In the second point of view, participants will look through the eye of a vortex column, reaching out far into space.

A few venues will have the ability to provide a full 360° view with a continuous ring of panels, although these will be the exceptions. However, the surround-sound audio in the room will be engineered to fully immerse the audience in a true 360° experience.

In the center of the room, guides help participants get organized using whatever mats and pillows they feel comfortable using. It is imperative that all participants find it easy and comfortable to move about the venue, as overcrowding will be counterproductive. All will need comfort and lots of elbow room.

As aiming participants settle in venues all across Tokyo, energizing participants elsewhere in the world are getting situated as well, but with very different staging, as shown in the Hawaii venue example below.

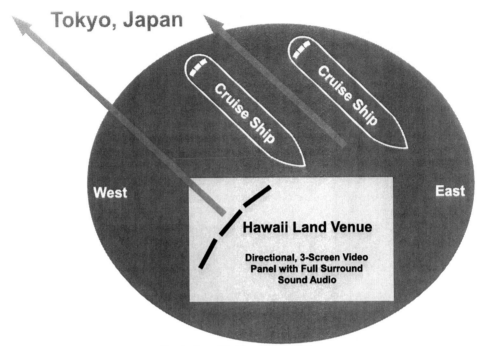

Illustration 65: Hawaii Land Venue

The principal goal of all energizer sites is to give a head-on view from the venue to Japan or wherever the projection zone for the intention vortex is situated. Curve-of-the-Earth, line-of-sight issues are not relevant as line-of-visualization views will be provided through event animations.

In Hawaii as well as everywhere else in the world outside of Tokyo, the staging of the energizing venues must orient local participants toward the projection zone in Japan. For this reason, optimization of floor space must support a line-of-visualization view aimed at the base of the intention vortex in Tokyo, Japan. Therefore, in this simulation, participants in Hawaii will face northwest.

Venues can be placed on land or at sea. In this case, calm weather allows cruise ships to anchor at sea, just offshore the big island of Hawaii. The sterns of their vessels will face northwest toward Japan, with large three-panel displays installed with surround-sound systems.

And now, as the clocks ticks down that last few seconds to 9 p.m. Tokyo time, everyone is ready to begin as anticipation on the island as well as on the cruise ship venues builds.

Tokyo: The Event Begins

The goal of this simulation is to create a small, but measurable change in the orbit of the second-largest moon of Planet X. Video cues for animations broadcast everywhere will be timed by event guides for the Imperial Palace venue in Tokyo.

The event begins...

9:00 p.m. JST: The Prime Minister of Japan opens the event and bids welcome to all participants worldwide and then hands off to popularizers of science for every language.

9:05 p.m. JST: For each language, a well-known popularizer of science (PoS) explains the aiming solution for the intention vortex and gives participants a visual description of the technical goal. This 10-minute presentation is commonly referred to as the Beta Stage Briefing.

The reason for that is when we are in the beta state, our minds are alert and working. This is the ideal state of mind for all participants worldwide when receiving the Beta Stage Briefing from their own native-language PoS.

At the close of the Beta Stage Briefing, each PoS hands off to a local event meditation guide.

09:15 p.m. JST: This three-part meditation immerses participants into the alpha state in preparation for the visualization. Whatever techniques, media, or tools are used, the goal is always the same: to help participants transition their minds from the active and working beta state to the relaxed and reflective alpha state during the first third of the meditation.

In this simulation, participants begin by lying comfortably on a floor mat. As relaxing natural sounds gently enfold them, they close their eyes and focus on their breathing. When the event guide feels the moment is right, participants begin with well-rehearsed alpha state immersion techniques.

In the second part of the meditation, the guide uses talking points prepared by the PoS to help participants contemplate essential key points made during the Beta Stage Briefing via meditative alpha state visualizations.

In the final third of the meditation, participants are slowly returned to a halfway state between the alpha and beta stages to clear their thoughts by imprinting the key points into their beta state memory, much like waking up just having had a dream so that you can write it down.

Participants are trained to view the memory imprinting through a life-affirming experience and during this last five minutes of the meditation, they relax their breathing as the three-panel video displays gracefully draw participants through colorful panoramas of nature —of the beauty that is our world.

Underneath this, the surround-sound experience can be as earthy as the distinctive and relaxing sounds of a Native American flute or celebratory like the "Four Seasons" by Vivaldi. What matters is that it works.

At 15 minutes, this is the longest segment in the event schedule; for remaining time, the only direction given to the participants will be five audio cues, each a simple chime tone.

Now it's show time.

09:30 p.m. JST — 1st Chime: By the first chime cue, most participants will already be resting with their eyes open. Their minds will be relaxed and clear for the task of starting the intention vortex.

At their own pace, participants stand upright with their feet apart and squarely face the center of the video panel before them. They see a live image of whatever is directly beyond them from a high point nearest the center of the projection zone.

At will, aimer participants turn their heads toward the left panel and begin a well-practiced movement one could describe as a tai-chi-style ballet. During this movement, they sweep energy with cupped hands—using broad, repetitive gentle sweeps from far right to far left. At the direction of the meditation guide, participants begin to see animation overlays appear in their view of the world about them.

They materialize as long wispy patches of golden, comforting vapors. This serves as a visual affirmation of the energy that aimer participants are infusing into the base of a new intention vortex. Over time, the animations become denser, with ever increasing counterclockwise velocity.

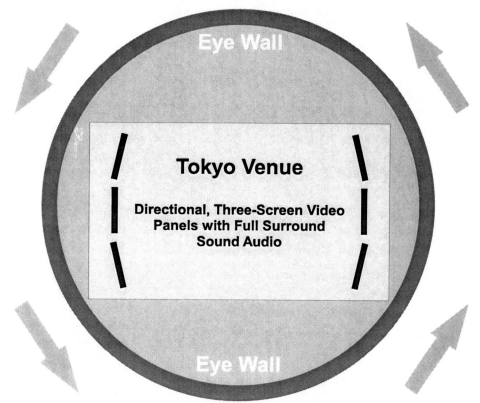

Illustration 66: Intention Vortex

As this is happening in Tokyo, participants in Hawaii see a completely different view on the center screens of their panels.

At the stern of their cruise ship, anchored at sea just off the big island of Hawaii, participants see a graphical view from far above their location—far enough that they see a wispy golden thread of light growing in the bottom center of the display panel. At the sound of the first chime, all participants begin their own well-rehearsed, repetitive movement.

With cupped hands touching at the fingertips, they hold their hands palm up at the waist. As they inhale, they raise their hands up along the body to shoulder height with a gentle scooping motion. Then in a continuation of the motion, they whip their hands strait out before them, visualizing their energy going to the base of the intention vortex forming in Tokyo.

As they do, the animation overlay of their energy trails outward as a translucent, golden light that follows the curve of the Earth to Japan.

Once that animation connects with the base of the intention vortex, the animation point of view travels forward until the view shown to participants in Hawaii is completely filled across all three video panels by the base of the intention vortex.

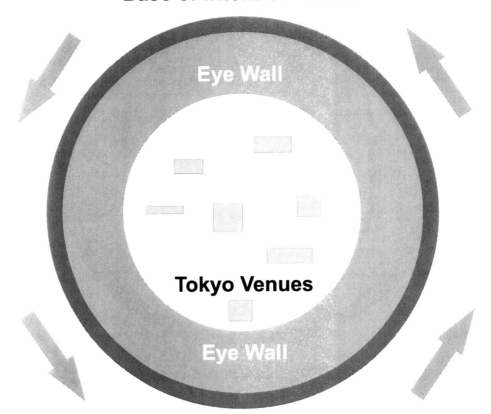

Illustration 67: Form the Base

No matter where energizer participants are in the world, this is how they will all merge with the aimer participants in Tokyo in preparation for the next step—raising the vortex column.

09:35 p.m. JST — 2nd Chime: With the second chime, all participants begin another series of movements designed to energize the vortex and raise its column skyward. In Tokyo, participants begin a counterclockwise circular swooping movement where they symbolically scoop energy from the base of the vortex and with an upward motion raise it up as a column.

In Tokyo, animation overlays on the display panels show the vortex forming a golden eye wall around the city, then rising up from all about them, and arcing skyward until a small circle of black space appears at the far edge of the vortex column. Though yet unseen, in the very center of that distant black circle in space is the aiming-solution target for the intention.

Meanwhile, elsewhere in the world, energizers will also see the same view upward that the aimer participants in Tokyo see. This is their cue to begin a repetitive movement where they both push and raise their energy toward Japan, conveying thoughts of love and support for the aimers.

09:40 p.m. JST — 3rd Chime: With the third chime, the panel view cross-fades from natural panoramas with animation overlays to a fully animated view of space. This effect is an integral part of participant training, so the transition is seen as being seamless and distraction free.

This cross-fade into a vertical, fully animated view of space is watched simultaneously by all participants worldwide as they as fly upward through the vortex column and out into space, straight toward the intention target designated by the science team.

During their training, participants learn that the cross-fade is also their cue for a flying motion. The moment they begin to fly through the eye of the vortex, they hold their arms out in front of them, and fly themselves upward.

As this spaceward flight slows, it finally comes to rest in a gentle hover in front of the second largest satellite of Planet X. It is at this moment that the magnitude of this effort is felt by those who are observing but not participating. After weeks of careful scientific calculations and planning, will it work?

As for the participants, reservations and hopeful doubts such as these will be the furthest things from their minds, as they next prepare themselves for the projection.

09:45 p.m. JST — 4th Chime: With the fourth chime, all participants sit on their floor mats and let their arms fall freely to their sides; they focus on breathing.

During the first part of this five-minute sequence, the hovering movement of the animation turns to a slow banking movement upward to the right. When it arrives at another gentle hover, the target (the second-largest satellite of Planet X) now appears in the lower left-hand side of the video panel.

From the target in the left pane of the video panel, a red, dashed line marches across the center pane and to the point of the intention goal in the upper right-hand side of the right pane.

When the line reaches the intention point of the aiming solution, it splits and wraps around an outline of the target at the intended destination point. Holding there for a few seconds, it fades out and the marching line animation continues to repeats itself, as participants impress the destination point into their thoughts.

At approximately one minute before the end of this segment, the dash line graphics fade out for the last time, leaving only the Planet X satellite in a side panel and empty space in the center and right panes of the video panel.

This is a visual cue to participants that the fifth and final chime is approaching and that this is the time to rise and stand upright with their legs apart and with their shoulders facing the center of the video panel.

Next comes the most important step of all. The one that transcends all others. The global projection of the intention itself.

09:50 p.m. JST — 5th Chime: When the chime rings for the last time, a moment of global resonance begins, a resonance that amplifies the projection of the intention a hundredfold, or perhaps even more. What powers this phenomenon is the harmonious collective of worldwide participants singing their own natural tones.

Each of us has a natural tone. It is seemingly effortless to sing once we find it. How do we know we've found it? When we sing our own natural tone from the heart and with natural emotion, it resonates throughout our bodies.

When we sing our natural tone beside others doing the same, we being to harmonize; the resonance is amplified by the sharing. Here is where the surround-sound systems of the event venues must reach their peak performance goals.

As participants sing in the various venues dotting the globe, their songs are broadcast live to regional channels where they are mixed and then rebroadcast live to the entire world. Not only is this a way to create a cohesive sense of connection as a species, it is a process that is literally driven by the growing sensation of that connection.

Here is where the human process drives the graphics in real time, through the following three phases of this segment.

At the start, participants begin singing their natural tones in a low voice, just above a whisper. As they do, the singing of other participants elsewhere in the world begins to harmonize with their own in a rich and resonant 360° audio experience.

At the right moment, the volume and intensity of this global resonance will reach its own natural critical mass, crossing a threshold one can only imagine within the sound itself. At this moment, the event guide in Tokyo signals the director to initiate a slow fade, where the Planet X satellite begins to dematerialize. As the target fades away, participants know to proportionally lower their voices along with the fade.

The director times the fade so that by the time the target has nearly dematerialized from sight, participants will have lowered their voices back to a level just above a whisper. Then, whatever remains of the target will blink out with a slight twinkle, giving the appearance that is has shifted into another dimension, out of phase with our own.

Once the participants see the target twinkle and blink out, they'll know to turn their attention to the destination point of the aiming solution. Participants will turn in their own time, and when the moment is right, the event guide signals the director.

With that, a red dash outline of the target appears again, as it did before, in the right pane of the video panel. Now the animation leaves with a bank to the right and then flies straight at the target destination. Taught to match the volume of their voices to the intensity of the animation, participants begin to sing out more, but are mindful to keep their greatest strength in reserve.

When the dashed outline of the target begins to dominate the center pane of the video panel, the animation slows to a stop and hovers for a moment until the dashed outline disappears, leaving nothing but a featureless star field behind.

This is the signal for participants to begin singing their own natural tone as loudly as is comfortable. As they do, they all take a slow step forward, toward the center of the video panel. They raise their arms and point their fingers at the center of the video panel, then begin singing their own natural note, with everything they've got to give.

With everyone singing boldly and joyously, the director fades up a barely recognizable image of the target. Faint at first, it grows slowly, but when the event guide cues the director, the target is outlined with a sparkle before it pops into view in full color.

The animation then hovers in place until the last minute, at which point the participants' view begins to retreat toward the earth, while Planet X and its second moon with its new orbit grow ever smaller. Meanwhile, the meditation guides begin easing participants out of the experience with warm and loving affirmations of success.

10:00 p.m. JST: It is done. In this simulation, the final message displayed on every venue video panel to officially close the intention event is lyrics from "Imagine," by John Lennon:

```
You may say that I'm a dreamer,
But I'm not the only one.
I hope someday you'll join us
And the world will live as one.
```

Illustration 68: Jonn Lennon

Now is the time to celebrate life with lots of feasting, hugging, kissing, dancing, music, and fireworks in celebration of our destiny as a more evolved, stellar species. This too will be an intention, a conscious thought to the universe that we're ready. And then we wait, but even as we do, many will come to a profound new understanding:

From all four corners of the world, we humans will have connected in a way never possible before and, in doing that, we will have glimpsed our future as a stellar species. That alone makes it worth the try.

Until the day comes when you feel yourself ready to become an Earth Pilgrim and to walk in the footsteps of Moses, look to the universe and say:

> "Today, I take personal responsibility for the survival and freedom of my species and I will do it alone if need be, and without any expectation of reward or recognition. This is my covenant with the Creator to which I bind my own life."

If you're not quite ready yet, that's OK. As a suggestion, this would be a good place to plant your bookmark for when you are ready, and not one day sooner. Wait for it and remember that all of this is really not about the tribulation, self-serving elites, or their minions.

It is about each one of us and how we choose to express and preserve our own free will. It is about our choices.

May we each choose wisely for one and all.

Catch you on the backside—Marshall

Alphabetical Index

Alphabetical Index

2004 Sumatra earthquake..................18
2012....6, 26, 28p., 32p., 38pp., 44, 46pp., 55, 58pp., 64, 74, 77, 79p., 104, 110, 112, 149, 184, 208
2013........................64p., 80, 82, 110, 122
2014........................79p., 82, 85
Aborigines..................10p., 75
Accelerant effect..................123, 129
Aesop's fables..................153
Albert Einstein..................74
Algae blooms..................69
American Stonehenge..................149
Aphelion..................46pp.
Arizona..........39, 141, 143, 158, 212, 215
Armageddon..................54
Aron..................7, 143pp.
Asymmetrical design..................31
Atlantis..................17, 67
Avebury....21, 23pp., 30pp., 40pp., 44pp., 49pp., 53, 55, 57, 63p., 72, 74, 79p., 91, 108, 110, 157, 207p., 238p.
Ayahuasca..................94
Betty..................127p., 216p.
Billy Mitchell..................225
Black Death..................3p., 10
Blue-green algae..................69
BP oil spill..................19, 106
Brown dwarf sun..................60
Bucket list..................184pp., 201
Captain James Cook..................13
Carl Sagan..................9
Catastrophism..........6pp., 79p., 103, 149
Catholic mystic..................56
Charles Hapgood..................74
Charon..................234
Chicxulub..................9, 188
Children of Men..................104
Chromosomes..................188p., 221
Church..................4, 8, 57
CME..................89, 207
CNSA..................244
Coast Redwood..................188
Colt .45 ACP..................199p.
Coma..................56
Comet.9, 17, 41, 43, 45pp., 54pp., 65, 82, 166p., 234
Comet hale-bopp..................45p., 61
Comets..................43, 54, 56, 234
Congress..................110
Consciousness. 31, 54, 66, 102, 119, 122, 137, 139p., 143, 150, 155, 159, 162pp., 169p., 176pp., 229, 241, 248
Contemplation..................152pp.
Corona..................51
Coronal mass ejection..................207
Coronal mass ejections..................89
Cosmos..................9, 120
Creationism..................6, 8p., 11p.
Creator.......78, 120p., 123, 126, 151, 155, 161p., 164p., 167pp., 172, 177pp., 189, 214, 218, 220p., 223, 228, 232, 258
Creators..................93
Cretaceous–Tertiary extinction event..188

Cro-Magnon..................................189p.
Crop Circle Connector....................27p.
Crop circles.................21pp., 31, 35, 215
Crossing the cusp......5, 51, 75, 91, 93, 98, 119, 121, 136, 147, 154, 159
Crystal Children..............................221
Cut to the Chase.....................23pp., 121
Cuvier..7, 9
Dark Ages...3
Darwin....................................6pp., 11
Darwinism...............................6pp., 11p.
Death with Dignity Act......................151
Deep Blue............185p., 213, 218, 236
Deep Impact....................................54
Déjà vu..135
Depression.....................................110
Destroyer........17pp., 38, 40p., 70pp., 129
Digital Millennium Copyright Act..........215
Disclosure videos.............58pp., 112, 215
Disinformation......22p., 28, 30, 58p., 103, 107, 112, 117, 214pp., 244
Disinformationalists...........................112
Disinformationists.............................217
Disney/Touchstone.............................23
DMCA..215p.
DNA......................................191, 221p.
DNIr4808n..................................59pp.
Dragon's tail...................53, 61, 63, 85
Dresden Codex..................................32
Earth........7pp., 14, 17p., 22p., 26, 28, 31, 38p., 45p., 50p., 54, 57, 61pp., 67, 69pp., 74p., 79, 81, 83, 85, 87pp., 96p., 111pp., 116, 123, 128, 159, 162, 166p., 169, 173, 175p., 178p., 187pp., 207, 214, 217, 221pp., 234, 238, 240, 257
Earth Pilgrim.............195, 201, 222, 257
Earth Pilgrims..............187, 189, 202
Earth Science..................................74

Earth-centric....................232p., 235p.
Earthquake..................188, 192, 206
Earthquakes....................18p., 81, 135
EarthSave48229pp., 237, 239, 241, 244p.
Echan Deravy.................................187
Ecliptic...............38, 40, 42, 48, 50, 234
Edgar Cayce....................14, 22, 50, 73
Egyptians.....14, 17, 66pp., 72p., 78, 93p., 129
Electromagnetic pulse........................111
Elites...4, 55, 84, 96pp., 103pp., 110, 112, 114, 150, 155, 166, 187, 219p., 258
Elohim..160
Emmerich...................................55, 104
EMP...69, 111
Entropy.................82, 116p., 120, 122
ESA...244
Europe........................3p., 64, 193, 242
Exodus........15, 17, 65, 68pp., 72, 78, 219
Extraterrestrial.....................8, 125, 176
Extraterrestrial entities.....................215
Extraterrestrials......................125, 214
Eyjafjallajökull volcano........................64
Failed star................................61, 65
False grandeur..............63, 66p., 83, 85
Family farms...................................105
Fear..5, 54, 58, 62, 71, 93pp., 103, 106p., 117pp., 121pp., 126p., 129, 134p., 137, 142, 147pp., 155, 159, 166pp., 177, 179, 187, 194, 206, 214, 229
Freddy Silva................................23pp.
Frightener..41
Garry Kasparov..............185, 213, 236
Genesis...............6, 13p., 20, 22, 76, 136
Georgia Guidestones.........................149
Global cataclysm.....10, 12, 14, 77pp., 85, 102, 119, 188, 195
Global catastrophe...6, 10p., 14p., 17, 22, 97, 231

Global human resonance event..........223
Global positioning satellite....................107
God...6, 17, 19, 76, 85, 93, 129, 158, 219, 246
Gods...71, 97
Goldilocks Zone..81
Gordon Lightfoot....................................141
GPS..107
Grand Canyon...........................50, 139p.
Great Deluge.....................14, 67, 72, 219
Greek philosophy..5
Green Revolution..............................104p.
H.G. Wells..5
Hammer............................30, 191pp., 201
Harmony. 10, 75p., 115pp., 120, 122, 155, 168, 173, 220, 222
Hawaii...............13, 245p., 248, 250, 253p.
Hebrew...............15, 17, 68p., 72, 78, 149
Hebrews..68
Hercolubus...17
History Channel..........................32, 104
Hitler..109
Holy Bible........................9, 13, 113
Homo Sapiens......................................218
Hopi..10p., 75
Immanuel Velikovsky...............................9
Indigo Children.....................................221
Infrared..58
Intention............8, 10, 28, 169, 179, 229p., 232pp., 243pp., 248, 250pp.
Intentions.................................229, 232p.
Islamist..106
JAXA...244pp., 248
Jewish..............................6, 219, 226
John Lennon...257
Junk DNA..221p.
Jupiter..................31, 43, 46, 55p., 58, 82
K-T impact...10

Kill zone.....48pp., 53, 56p., 63, 73, 80, 84
Knowing...16, 102pp., 138, 148, 159, 161, 165, 173, 217
Kozai mechanism..................................44p.
Law of Entropy...82
Law of Thermodynamics.........82, 91, 116
Lee-Enfield..204pp.
Libra..6, 49
Lifeworkers........115p., 118pp., 123p., 127, 129, 131, 139p.
Lightworkers....115p., 119, 124, 126, 137, 157
Lithosphere.............................20, 74, 91
Lobal cataclysm......................................7p.
Loma Prieta earthquake........................84
Lost Book of Nostradamus....................32
Louis L'Amour.......................................216
Louis Pasteur..134
Love.....75, 93, 97, 115, 117pp., 126, 129, 139, 145p., 150, 159, 161, 164p., 167pp., 175pp., 179p., 207, 209, 220, 232, 247, 255
M. Night Shyamalan................................23
Magnetosphere.........................89, 221p.
Mark Fussell..27
Mars.....................9, 23, 28, 31, 50, 82
Masada...226p.
Mauser..204p.
Mayan calendar......................................32
McCarthyism...9
Meek.......6, 112pp., 121, 123p., 129, 131, 134, 137, 155, 216p.
Meensen..40
Mercury......................................31, 33, 51
Mesopotamian..9
Microcystin..69p.
Millennium bug..................................107pp.
Minions. .84, 96p., 103p., 112, 150p., 155, 218p., 258

Moses.....65, 69, 72, 158, 219p., 222, 257
Mosin–Nagant..................204p.
Mother Earth....................232
Mother Shipton.........61, 63, 66pp., 83, 85
Mount Graham...................38
Napoleon.......................109
NASA................54, 236, 244
National Geographic.............22
Natural Flow of Consciousness.....119pp., 123, 129, 133p., 147p., 150p., 155
Neanderthals.............189p., 214
Near Earth object..............232
Near Earth objects.............233
Near Earth Orbit................54
Near-death.........137, 140, 144, 154p.
NEO...........................7, 54
Neocatastrophism.................7
Neptune..........................31
NGC............................22p.
Nibiru.....9, 17, 21, 27, 30, 41p., 59p., 112
NibiruShock2012...........59pp., 112
Nietzsche......................218
Noah...8, 13p., 20, 22, 67, 75pp., 97, 123, 129, 196
Noah's flood.........8, 13p., 20, 67, 72, 219
Nostradamus....................32, 45
Occam's hammer..................30
Oh My Johnny..................146
On the Origin of Species.........7
Ophiuchus....................32, 49
Out-of-body........137pp., 143, 154p., 212
Out-of-phase.........137, 140, 142p., 146
Patty Greer....................25p., 37
Pearl Harbor...................225
Perihelion.............19, 46pp., 50, 64
Pharaoh of Exodus............69, 219
Phoenix......................4, 138
Phosphorus.....................69

Photojournalist.............139, 143
Plague..................3, 17, 65, 69
Planet X.....17, 20p., 24, 26pp., 32, 38pp., 44pp., 53pp., 69, 72pp., 79p., 82pp., 86pp., 90p., 122, 128p., 171, 219, 230, 233p., 236, 241p., 245p., 251, 255pp.
Planet X and The Kolbrin Bible Connection....................67
Planet X Forecast and 2012 Survival Guide............28p., 124, 207
Pluto.......................43pp., 234
PoE..............82, 118pp., 129
Point of equilibrium.........82, 85, 118, 140
Pole shift....14, 18pp., 32, 50p., 72pp., 77, 79p., 85, 90p., 94, 96, 110, 123p., 127pp., 167, 172p., 175, 219p.
Prescott...........141, 143, 158, 160, 212
Psychoactive effect............94p.
Quiescence.......7p., 10, 15, 18, 79, 85
Recession......................110
Red Comet...................9, 17, 41
Renaissance......................4
Romans.........................227
Roscosmos......................244
Russia....................9, 106, 109
Russian........................149
Russians.......................109
Sagittarius.....................49
Saturn..........................31
Schreibersite...................69
Science................4p., 8p., 11, 23
Science Channel...............104
Scopes...........................6
Scorpio.........................49
Serapis Bey..................216p.
Sesame Street.................240
Shakespeare.....................85
Shared-death........137, 143, 145p., 151

Signs..23, 172
Sisuda and Hanok................14, 17, 20, 75
Sleeping prophet...............................14, 73
Solar. .9, 17, 20, 22, 38, 46, 49pp., 53, 58, 73, 80pp., 88, 91, 108, 110p., 122, 166, 171, 175, 207p., 234
Solar Code...187
Solar radiation....................88, 207, 221p.
Solar system.38pp., 42, 44pp., 48, 50, 55, 66p., 74, 159, 166p., 171, 175
South Pacific...147
South pole......................................58, 166
South Pole telescope...................58, 215
SPT...58
Starry Night over the Rhone...............227
Stellar species 5, 97, 155, 157, 218p., 257
Stuart Dike..27
Sun. .18pp., 32pp., 38, 43p., 46pp., 55pp., 60p., 64, 71, 81p., 87p., 96, 158p., 162, 166, 171, 175, 179, 221, 247
Supernormal......134p., 152, 207p., 214p., 217, 222, 232p.
Suzanne Taylor....................................25p.
Symmetrical design................................31
T. S. Eliot..241
Ten Plagues of Exodus......17, 51, 67, 69, 219
Tennessee..6
The Book of Eli....................................104
The Bronzebook......................................69
The Day After Tomorrow.....................104
The Earth's Shifting Crust.....................74
The enlightenment......97, 131, 133, 154p.
The Great Winnowing..72p., 93pp., 116p., 121pp., 129, 131, 134, 136, 148, 154p., 159, 187, 217, 219
The Hollow Men.................................241p.
The Kolbrin Bible.....14p., 17p., 65, 67pp., 75, 78, 93p., 129, 219

The minestrone........................35, 37, 43
The Poseidon Adventure.......................84
The Road...104
The Time Machine....................................5
Thomas Henry Huxley..............................7
Tokyo...................237pp., 245pp., 253pp.
Torah...6, 14
Touchstones.................................129, 131
Tramlines..36p.
Trigger event...........8, 10, 21, 50p., 230p.
Tsunamis..135
Two suns.....................49, 63, 66, 135
Uranus..31
Ursula Southeil.......................................61
Vatican Advanced Technology Telescope 38
Venus........................9, 31, 33, 46, 51
Veronica Lueken.............................56p., 63
Vincent van Gogh.................................227
Virgin Mary..57
Visualization. 127p., 186p., 190p., 194pp., 212, 238p., 251
Visualizations............126p., 194, 208, 239
Vladimir Putin..106
Waning Crescent Moon..........................49
Weather Channel..................................104
Worlds in Collision....................................9
Wormwood..............................17, 41, 73
Wreck of the Edmund Fitzgerald.........140
WWI................105p., 146, 200, 204p., 242
WWII.................................105, 146, 225
Y2K..107pp.
Yoshihide Kozai.......................................44
YouTube.com........................58pp., 112
Yovy Suarez Jimenez.........................198p.
Zodiac..38
Zones of control.................................195p.

CPSIA information can be obtained at www.ICGtesting.com
Printed in the USA
LVOW11s0503140813

347573LV00005B/80/P